Animals Who Own Us

Keep telling your animal stories!

David B Hazelwood

2/26/22

David B Hazelwood

Also by David B Hazelwood

Lists for Ainsley

Cortner Mill Cookbook

Miss Lizzie's Cookbook

Monday Morning Rose

Burgoo: It's a Kentucky Thing

Cooking Southern:
Recipes and Their History

Animals Who Own Us

David B. Hazelwood

WUFU3.com Publishing
Normandy, Tennessee

Cover design by Anne Craig, Craig Communications
Cover photo by Ted Wizer

Published by:
WUFU3.com
Normandy, Tennessee

ISBN 978-1-734-8305-1-4

Contents

Introduction

Be careful what you own; it will eventually own you. You don't own an animal. It just lets you think you do. At first, you're putting them in their proper place, making them look the way you want, and training them to obey your wishes, but soon they are telling you what to do and when to do it. They punish you when you don't.

"I need to go out," the dog says. "Open the door now. Take me for a walk." I ask her, "Do you want to come inside?" She gives me a blank stare as if to say, "No, I'm not ready now. Feed me. My foot hurts, take me to the vet. Buy me some more food. Buy me a new collar." She owns me. That's why people see our Golden Retriever, Lane, and ask, "Is this your dog?" I reply, "No, I'm hers. She's my master. She owns this place and I work for her. She tells me what to do and when to do it. She's the CCO, Chief Canine Officer." I wouldn't have it any other way.

I've had animals constantly all of my life, usually several at the same time. When one would leave, I would start looking for another. I must have been a slave in a previous life. I've been owned by cows, pigs, chickens, horses, ponies, ducks, peacocks, turkeys, rabbits, parakeets, goldfish, guppies, Cocker Spaniels, Beagles, Irish Setters, Great Danes, Springer Spaniels, Golden Retrievers, German Shepherds, Strooches (half stray, half pooch), strays, Calicos, Tabbies, Manx, ferals, drop offs, loaners, Angus, Herefords, Jerseys, Holsteins, Charolais, Simmentals, Limousins, Chianinas, Black Baldies, and Beefalo. (In case you didn't know, those last ten are all cows.) No sheep, goats, emus, or llamas yet, but life isn't over. I've always thought I wanted a monkey, but never had to say, "I'm owned by a monkey."

Do you need to be told what to do? Go to the pet shop and pick out a new owner. Need something to do twice a day, every day? Get a dairy cow or goat to milk. Live in an apartment and never seem to get out? Need exercise? Get a dog that can't pee or go for a walk without your help.

Having a thumb is the only evolutionary advantage I have over all the animals that have owned me. I'm using it to write their stories. They told me to do it. I'm sure this is volume one because there will always be more animals to own me and they will all have stories to tell. About the time I think I've seen all of the situations animals can get themselves into, I come home and find my dog with a trash can stuck on its head.

I didn't start out to write all stories about the animals in my life. I started writing for an audience of one, my daughter, Whitney. One holiday I told a story about one of the animals I had while growing up. Whitney said she had never heard that story before and it made me think there were probably a lot of my stories she had never heard. I've always been a story teller and avid reader of a book a week, but had never written anything except articles for professional magazines. So, I took a year off from reading and just wrote.

After sharing several of the stories in this book with my reading and writing mentor, Mary Driscoll, a retired English professor, she suggested I send a couple of the stories to a national magazine for their consideration. Their editor was considerate and kind, but said, "Sorry, we only publish true stories." Folks, like I told him, everything in these stories is true. I couldn't make up these things. When they happened, I was as surprised as you will be when you read them. Since you are reading this book, my audience is now two!

David B Hazelwood

A Grandfather, A Boy, A Pony

Quite a triangle! I was nine years old and for several years had gone to my grandparents to stay a week or two during the summer or other school holidays. When they would come to our house, I would pack my suitcase (probably a pair of jeans, a shirt, socks, and maybe a pair of underwear for a week) hoping, expecting to go home with them. I usually did.

In the first, second, and third grades my day dreams were about ponies. One could buy one from the Sears-Roebuck catalog for $150 complete with saddle and bridle. I would look at the pages of saddles over and over again to decide which I would buy someday. At school, when I should have probably been doing something else, I would dream of pony races between my cousins and friends at my Uncle Ronnie and Aunt Thelma's farm. They had a whole field of ponies that were used to pull coal cars in the mines. My favorite was Dan, even though, or maybe, because he was blind in one eye. He was surely the fastest runner. In the races of my dreams I would claim Dan as my mount and would always win. In real life one of my best friends in elementary school was Paula Lynn. In my eyes her best attribute was that she had a Quarter horse that she took to horse shows for barrel racing.

On the first day of one of my summer visits to my grandparents, my grandfather told me to come with him. He had a surprise for me. Being the practical type, my grandfather wasn't much into surprises. It was a surprise just to hear him say he had a surprise for me. I couldn't imagine what it could be.

As we rode along in his pickup truck, there wasn't any conversation to give me any clues about where we were going or what the surprise might be. After about three miles we got to the little village of Hebbardsville and turned down Alvis Ferry Road that went by my other grandmother's house. But the surprise didn't have anything to do with her and we drove past. After another quarter mile we turned into a driveway and headed up the hill to a brick house. I didn't know who lived there and had never been there, even though it was close to my grandmother's house. I still didn't have any clues to the surprise.

After a few words of greeting between Paul Mauzy, who lived there, and my grandfather, we turned and walked toward the barn. Mr. Mauzy and my grandfather talked about the weather and crops as I shuffled along behind them. About the time we got to the barn, out of the darkness of the barn walked a girl about my age, leading a pinto pony already saddled and ready to ride. She introduced herself as Molly, but all of my attention was fixed on the pony. Even though I liked what I was seeing, I still had no idea what all of this meant. Was I coming here for a pony ride? Was that the surprise? It was a good one if it was. They kept talking about how their tobacco crop was coming along and paid little attention to the arrival of the pony. No one invited me to mount up, so I just stood beside this wonderful animal and stroked the silky, smooth hair of its neck. Molly told me the pony's name was Flicka. Not a very original name since only thousands of boys and girls had named their horse or pony Flicka after the horse in the book, My Friend Flicka.

The surprise wasn't that I was being brought here to ride Flicka this afternoon. My grandfather began telling me how he had arranged for us to borrow Flicka and take her back to his farm, where I could ride her all week long! I couldn't believe my ears. I was so excited that time seemed to speed up. All of a sudden there were so many things to take in. This was no daydream. It was real.

Nothing had been said about how we were going to get Flicka back to my grandfather's farm. He didn't have a horse trailer and there weren't any racks on the back of his pickup for hauling livestock. It was just one pony and I knew my grandfather wouldn't have hired Fatty Joe Butler to bring the big truck that usually came when hogs or cows were hauled to the stockyard. Since no one seemed to have thought about hauling the pony, I asked how we were going to get her back to the farm. My grandfather announced I was going to ride her back. That sounded even better. There wouldn't be any waiting for a pony ride or any wasted time. The riding would begin immediately and last all week.

And begin immediately it did. Molly stood beside Flicka and held the reins as I awkwardly climbed into the saddle. I was feeling on top of the world. Molly didn't seem to mind at all that I was taking her pony for a week. I couldn't imagine it, but maybe having a pony to ride anytime you wanted had gotten to be old to her. It really didn't matter to me; I was going to ride every moment, all week long. Nothing would get me to stop riding. Up until this point, my riding career had consisted of flying like the wind in my daydreams and going around in circles on a pony tethered to a carousel at the park. Neither was the real thing, but I was about to experience the real thing. My first real ride was going to be three miles back to the farm. It would be like a real cowboy on the Santa Fe Trail, only without the cow herd. So, full of wonder and some nervousness from inexperience and having to perform in front of this girl, I headed out of the barn lot and down the driveway. It was just me and Flicka.

Flicka seemed to be on familiar territory and knew where she was going, so it didn't take much reining or encouragement from me. My grandfather trailed along behind in his pickup. The first leg of the trip went by quickly and uneventful as I got used to the feel of the saddle and the reins. We walked the first quarter mile and as we came to my grandmother's house, she came out the front door and onto the porch. I gave her a big wave and reined Flicka into her front yard. My two sets of grandparents didn't talk to each other often, but they must have talked about the pony because Grandmother didn't just happen to see me riding by and come out to wave. We all admired the pony and talked about what a wonderful week of riding it would be. She knew I wouldn't be sharing any of that week with her. There wasn't anything at her house that held the charm of a pony, not even her chess pie, cinnamon toast, apples with red hots, or a walk to Bryant's Grocery for a cup of pineapple sherbet or peanut butter log candy. This was a real, live pony.

The afternoon sun was moving lower and the heat of the day was already easing up. There was still a lot of the ride ahead of us so we couldn't talk long, even though Flicka seemed content to just graze on the lawn while we talked. We said our good byes and made our way back to the road. In another fifty yards we were at Bryant's Grocery, made a left turn onto Highway 54, and headed into the afternoon sun. Even though Highway 54 was a state highway that ran between Henderson and Owensboro, where I lived, it didn't have a lot of traffic. Only about every five minutes would a car come by. Flicka and I walked in the grassy roadside as my grandfather came along slowly behind us in the pickup. The road in front of us was long and straight as it moved through the flat creek bottoms at the edge of Hebbardsville. Even though just riding on the back of this pony seemed plenty exciting to me, walking along must have been too slow for my grandfather. He pulled up beside me and called across the cab through the open window, "See if you can make her trot." He wasn't testing my riding skills. He figured that at the rate we were going it would be dark before we got to the farm. As adventurous as this was, none of us wanted to be out here along the road with a nine-year-old on the back of a pony in the dark.

I shook the reins like I had seen the cowboys on TV do to get their horses to go faster. Either Flicka didn't know what it meant, I wasn't doing it right, or she wasn't interested in going any faster. So, we just kept walking along slowly. I talked to Flicka like she was a playmate. I told her how much I liked her and what all we were going to do together during the week. She gave a snort and shook her head, as if to say she couldn't wait either. I asked her if she had ever been in any pony races. She just trudged along. Once again, my grandfather pulled up alongside. This time, rather than suggesting I try to make Flicka trot, he asked me why I couldn't do it. His words stung. While I was disappointed that my first attempt at changing our gait wasn't successful, I wasn't sure what to do about it. He suggested I give Flicka a kick in the ribs with my heels. I didn't want to kick my beloved pony for any reason. I was somewhat fearful that I might get more reaction and speed than I wanted. I wasn't quite ready to fly like the wind on the pony of my daydreams. So, I nudged her gently with my heels. There was no response, and we continued to walk along. "Kick harder," my grandfather said. I tried kicking harder, but we still just walked along through the bottoms and up the hill past Priest Hazelwood's house. He wasn't a priest. That was just his name. He was my grandfather's second cousin twice removed or something like that and owned the bottom land we had been walking past.

By the time we got to the top of the hill, nearly thirty minutes had passed since we left the Mauzy's barn, and we had only gone about a mile. I could tell my grandfather was getting impatient. We came to a small side road that cut through the bank of the hill, and the roadside narrowed so much that there wasn't enough room between road and the ditch for me to ride. My grandfather suggested I take Flicka up onto the bank where I could ride in the field alongside the road. The bank wasn't but about three feet high, but it looked like the Grand Canyon to Flicka and me. She didn't want to go up it. I was about to realize that all along this ride Flicka had been in control. She had controlled the speed, and now she was going to control the direction. Even though my rein shaking and rib kicking wasn't effective in getting her to trot, I tried them again to get her to go up the bank. We would head toward the bank, and just before we started up, Flicka would turn and go along side it. Several times I circled her around and made another attempt, but each time she would turn away when we got to the edge of the bank. My grandfather shouted out, "Use your reins and whip her on the neck." No boy likes the thought of a whipping and I didn't like the idea of whipping her soft neck. But my reputation as a horseman was on the line, and I slapped her neck several times with the ends of the reins. It only seemed to agitate her and she just shook her head unmoved. Since I didn't like the idea of whipping her to start with, I quit and just sat there staring at the bank and thinking my dreams of becoming a rider had turned into a nightmare. I had failed to do what my grandfather expected and couldn't imagine anything but a week of riding this pony was lost.

So much time had passed that my two uncles, Arnold and Larry, must have thought something had gone wrong and came looking for us. After some discussion, it was decided that Larry should ride the pony the rest of the way home. Larry was ten years older than me and 100 pounds heavier, probably too big to be riding the pony, but there didn't seem to be another alternative. Larry got onto Flicka and it looked like he was bigger than she, but off they went. I rode in the pickup as my grandfather drove along behind Larry and Flicka. Even though Larry was older than me, it hurt my pride to see him doing what I dreamed about, but couldn't do when I had the chance. To make matters worse, he slapped Flicka on the neck with the reins and gave her a kick in the ribs and off they went into a trot. I wished I was the one going that fast on Flicka, but I was riding in the truck and Larry was the one getting the ride. Flicka's feet moved so fast, they were a blur. And with movement of each hoof there was a bounce transferred to Larry's seat. The bounces came so fast with the pounding of each hoof that Larry didn't have time to come back down on the saddle before he was being sent up again. His body seemed to be going eight directions at the same time. Normally he could have cushioned the bounce with his feet in the stirrups, but his legs were too long to fit and he had to stay on by wrapping his legs around Flicka's belly. He couldn't hold on with his legs so he just kept bouncing. A big smile came on his face which made me think he was having a great time riding this pony. But the smile was the beginning of a laugh and the laugh was not from having fun but at the out of control situation he was in. He knew he couldn't keep this up for two miles and probably had visions of flying off at any second. He either had to slow down or go faster. Since slowing down seemed to defeat the purpose of him riding instead of me, going faster was the only option, and he kicked Flicka into a canter.

After a few strides of the easier gait, Flicka decided she couldn't carry this big guy at such a pace. A slap of the reins not only failed to make her go faster, it brought her to a stop. She had had enough of this. Larry went into a simultaneous whipping and kicking mode which sent Flicka into a stopping and bucking mode. She wasn't going to carry this big guy a step further in any gait—walk, trot, or canter. Larry decided rather than getting bucked off, he would just get off. Flicka shook all over like a wet dog shaking off water. She was glad to be rid of Larry and he seemed to be glad to be standing on ground that wasn't bouncing him up and down. My grandfather and Uncle Arnold came over just laughing at Larry. "What's wrong? Can't ride a pony?" they joked.

Now what were we going to do? If the older, more experienced Larry couldn't ride Flicka home and I couldn't make her go faster than a walk, how would we get her home before dark? Arnold said we should put her in the truck and haul her home. I couldn't imagine such a thing. There were no racks on the sides of the truck bed to keep her from jumping or falling out. Besides, how would we get her into the truck? There wasn't a loading chute around. Once again, I began to doubt if we would ever get her to the farm for me to have a week of riding.

My grandfather saw a bank on the other side of the road and said we could load her from the bank. He backed the truck into the bank, and the bed of the truck was just level with the bank. Larry led Flicka up the bank and toward the truck when she put both feet in front of her, braced herself, and came to a stop. She didn't like the idea of getting onto the truck. Larry pulled on the bridle reins, but all that moved was her head. My grandfather and Arnold got behind Flicka to push while Larry pulled. I had always been warned about getting behind a pony, and I just knew they were going to get kicked. But Flicka gave into the pushing and pulling and in an instant bolted onto the truck. With Larry on one side of Flicka and Arnold on the other to reassure her and hold her in place, my grandfather and I climbed into the cab of the truck and eased the truck onto the road toward home. I watched Flicka through the back window and we crept along the last mile and a half. It was a slow, easy truck ride, but faster than my riding gait and smoother than Larry's.

As each day of the week went by, Flicka gave me lessons on riding and pony behavior. I learned what she liked and didn't like. I learned what she would do and wouldn't do. I learned how to trot without bouncing my brains out. And best of all, I learned how to race like the wind, leaving all the other ponies in our dust, just like I had dreamed about.

Nunn Better Chicken Feed

What was the name of that chicken feed? I've got so many things to remember and the TV announcer said I should tell him the name of the chicken feed when he asked. Where did they put that sack of feed? I looked around the TV studio. This place didn't look anything like I expected. It was more like being in a garage with bright lights, than a place to do TV. But when you're eight years old, a lot of places you go are new to you. Being in a TV studio wasn't new only to me. In 1955, not many people had been on TV. I was going to be the first in my family to be seen on TV. I was so excited that my mind couldn't keep everything straight. What could my mom say to keep me from being nervous? She hadn't been on TV.

Everything here was so big. The cameras must have weighed 200 pounds. All of the set props were much bigger than I had seen on TV. The big people walking around everywhere made me feel I was about to be stepped on. I finally spied the feed sack leaning against a counter. What a relief. Now I can find out the name of that chicken feed again. It was a 100-pound sack made of a floral cotton print. When the feed was all used, a farmer's wife could use the material from the sack to make a dress or blouse.

Nunn-Better. Now that's a catchy name for the Nunn family to give its chicken feed. They were the reason I was about to become a TV star. For now, I would just stay out of the way by sitting on this feed sack.

The Nunn Milling Company in Henderson was the local sponsor of a weekly TV show. During the commercial breaks, a man from the TV station would stand in front of the camera and tell everyone in the tri-state area that there was "none better than Nunn-Better" feed. Tonight, I would be standing beside him. I had been coached by parents, aunts and uncles, teachers and friends. I was ready to answer any question he asked. I was ready to be on TV!

This all started a few weeks before, when during the Nunn Better feed commercial they announced a contest. To enter, contestants had to be ten years old or younger and write a letter to the TV station telling why they wanted to win 50 baby chickens and a 100-pound sack of Nunn Better feed. Now I was sitting here on part of my prize with the rest of it cheeping in the background. I must have really pulled some heart strings in that letter to win. I sure didn't outline in it what I planned to do with the chickens, if I won. It didn't matter that we lived in town, in a new subdivision, with nothing on our unfenced lot but the house. Where would I keep them? What would I do with 50 baby chickens after they became roosters and hens? None of that mattered. Just remember when he asks what I will feed them, say, "Nunn Better feed."

As commercial time came closer, the pace of everything quickened. People scurried around. One man came for my feed sack chair. Another led me into the bright lights, beside the commercial announcer and in front of the camera. Time was both standing still and moving faster. My heart pounded. My breathing stopped and so did my brain. What was the name of that feed? I couldn't forget. He said be ready when he asked.

Finally, something familiar. The announcer started talking to the camera about the wonders of Nunn Better feed. I was on TV and my moment of stardom was about to come. I was ready. He looked down and asked me a question, but it wasn't the question that my Nunn Better answer fit. It just required a simple "Yes" or "No" response. My mind began racing as fast as my heart. I wasn't ready to say "Yes or No". I was ready to say "Nunn Better." Now what? I wasn't ready for adlibbing. What did he ask? What should I say? All I could remember was the advice from my teacher, Mrs. Miller. Be polite. So, I blurted out my first TV broadcast… "Yes, ma'am!"

The announcer looked at me and gave an all-knowing smile. He looked back to the camera and continued his Nunn Better message, never to ask the question I was prepared to answer. I knew then my TV career was done for. Now I had to go back and face all my family and friends with the thought that mommy's boy had said "yes, ma'am" instead of "yes, sir" to a man.

Walking out of the bright lights, time returned to its normal pace. My heart had stopped pounding and I could breathe again. But as we loaded the feed and chickens into the car and I accepted my final congratulations, my mind moved ahead to the next stop. A room full of aunts and uncles had gathered at my grandparents' around a little black and white TV to see and hear my one minute of fame. As soon as I came through the door, I knew the first words I would hear. "Yes, ma'am," Uncle Larry would say. From then on, it would be the two words guaranteed to get a laugh from everyone at the family gatherings. My grandmother would call off the teasing by telling them I was just excited. And after a few more laughs from everyone, my granddad would help me exit gracefully by taking me away for the more practical matters of putting the baby chickens in their new home at his brooder house. Dad had already arranged for granddad to keep the chickens for me. I would get to go to the farm to stay a week or so several times during the summer. I would get to fulfill my responsibility of tending the flock by feeding and watering them every day during the visit.

I don't know if it had been pre-arranged or not, but at the end of the summer, Granddad approached me with a deal on the chickens. I had spent weeks watching those chickens go from chicks to fryers. Would I have to kill, dress, and sell fresh chickens on the roadside? First, Granddad told me how much the feed bill was. I learned the Nunn Better folks had given me just enough feed to get those chicks started. Then he told me about the medicine he gave them when they got some disease that made their toes crooked. I was beginning to think I would end up in the hole on these chickens I had won, when he proposed giving me a $20 bill. He would keep the chickens and we would call it even. He asked if that sounded like a good deal to me. With relief and pride in a deal well struck, I stood tall and said, "Yes, sir!"

Adventures with Jack

My first experiences around equine were at my grandparents' farm with their three mules. I watched them work, inspected their harness, learned their commands, felt their noses, and even got to ride a few times. They weren't the ponies I dreamed about but made a good substitute at the time.

By the late fifties horse and mule power was at its end. Only a few niches gave them a productive place. They still had a place in cultivating small gardens until rotary tillers were developed. And like on my grandparents' Kentucky farm, they were used for planting and cultivating tobacco until they were replaced in 1960 by the Farmall Cub and Super A tractors.

My grandfather had three big Missouri mules named Jack, Jim, and Dick. Jack and Jim were the ones who did most of the pulling jobs. Dick was more energetic than the other two. This made him less reliable and more likely to not make a straight row or to step on a row of plants when cultivating. I never saw him being worked, so he must have been a backup in case one of the others was sick or injured.

As soon as school ended in May, I wanted to head to my grandparents' farm to be part of planting the tobacco. Just being on the farm when the mules were being used gave me opportunities to learn the basics. I was fascinated at feeding time when they came running into the barn after being called. I learned to yell "Wuuuuuuu!" like my grandfather. They would each go into their designated stall, as if their name was above the door, and wait patiently for the ears of corn I put into their feed troughs.

It was a big thrill when my grandfather offered to let me a ride astride one of the mules as we went to or from the fields. I got to hold the reins, even though my grandfather was really doing the driving with the lines. I listened to my grandfather call out "Gee" to make them go right and "Haw" to make them go left. Of course, I already knew "Giddy up" and "Whoa."

At tobacco planting time I would walk behind or alongside the setter as my grandfather drove the mules slowly and straight down the field with my uncles, Arnold and Larry, riding the setter and putting the plants into the ground. I saw how important it was to maintain a steady pace. If the mules went too fast, my uncles would yell out "Slow down!" because they weren't able to put the plants into the ground fast enough. Making the turn at the end of a row, they looked back at their work. If the row was crooked, they teased my grandfather about driving like he had fallen asleep or was drunk.

Walking alongside the setter's slow-moving wheel, I would let it run over my foot. With a combination of the setter being a light piece of equipment and the ground being disked into a soft mash, I didn't feel a thing.

After several rounds, the setter's water barrel would need to be refilled, which gave everyone a break. I used the break to inspect the mules. I looked at their harness to see how it was all connected and what functions each part performed. I rubbed their velvety noses, while smelling their hot breaths. Patting their lathered coats wasn't very appealing after the first time.

On my visit in mid-summer, it was time to cultivate the tobacco and I got to walk along behind my uncles or grandfather, as a mule pulled a three-point cultivator. Years later I discovered the job wasn't as easy as it looked. In addition to driving the mule, I had to control the cultivator. The mule was actually the easiest part because he knew what to do. But it seemed to defy logic that to make the cultivator move away from the row I should move the handles toward the row. I plowed up quite a few stretches of row before I figured it out.

Another mule job was more fun for me than cultivating. If the tobacco had been cultivated while the soil was too wet, clods of dirt were created. To break them up my grandfather made a clod buster. He made a box our of 2 X 6 lumber that was the width to go down between the rows of tobacco. He took handles off a cultivator and attached them to the box. As the mule pulled the clod buster between the rows, it pulverized the clods. The fun part for me was getting to ride in the box. I had quite a view moving down the rows at ground level. My eyes were on level with the mule's hocks, and I could see their hooves press into the dirt two-feet away. Being that close to the hooves made me remember the family story about my Uncle Hugh being kicked by a mule when he was my age. He was knocked unconscious and lost a tooth.

My Uncle Larry knew I was anxious to have a mule ride and caught Jack one afternoon. Jack was my favorite because he was the most docile. Larry put the bridle with blinders onto Jack. There wasn't a saddle so we rode bareback. All I had to hold onto was Jack's mane, which I didn't need to do when we were just moving at a walk through the barn lot. Larry even let me hold the reins, and guide Jack to the pasture.

Going through the pasture gate Larry took the reins and gave Jack a kick in the ribs. Off we went at a trot. I was bouncing up-and-down and from side-to-side. Holding onto Jack's mane wasn't giving me much control, and I knew I would fall off any minute. Larry held onto me and kept me in my place, but the trotting was making his ride unsteady too, so he gave Jack another kick in the ribs and we surged into a gallop. Jack headed straight for a group of cows and made them scatter. As we passed close to one, Jack barred his teeth and reached to take a bite of the cow. Even though they shared the same pasture, I didn't realize mules and cows weren't on good terms. As the cows ran away, I knew my grandfather wouldn't be happy knowing we used the mule to chase the cows. The galloping ended as we resumed a walk, but the exhilaration continued the rest of the afternoon.

As the months passed, my feeling experienced with the mules grew. I'm sure my feelings as an eight-year old exceeded my true experience. During Christmas vacation I got to visit my grandparents for a week. It was tobacco stripping time on the farm so there wasn't any work for the mules, but at least they were still around to be fed and admired.

My job to help with the tobacco stripping was to carry the stalks that had been stripped of their leaves out of the garage and throw them onto a wagon or into a high pile. They would later be scattered onto a field where they would rot and add organic matter back into the soil. The rate of stalks being stripped was slow enough to allow time between trips for me to fool around. Sometimes I climbed to the top of the twenty-foot high pile of stalks just for the view.

One day from my perch on the stalk pile I saw Jack standing beside the fence near the garage. I went over to him and climbed onto the top of the fence and patted him on the neck. He never moved at all, except for a twitch of his long ears. I rubbed his head and neck for about ten minutes. His broad back looked so inviting that I slipped off the fence and onto his back. I was clueless to the danger I had put myself into by getting onto an un-haltered, untied mule.

I hadn't been sitting there talking to Jack very long until Uncle Arnold came out of the garage with an armful of stalks that had accumulated while I was on my adventure. I can only imagine now what he must have been thinking. After quickly tossing the stalks onto the pile, he came straight to me and gently lifted me off Jack's back, over the fence, and onto the ground. All he said was "I don't think that was a very good idea." No one else ever said anything to me about it, so he must have kept it to himself.

The last time I remember riding Jack was the next summer just before I turned ten and saw myself as "all grown up." I had been left at the house to spend the day with my grandmother. That usually involved something like working in the basement using the wringer washing machine to wash and rinse clothes or helping her in the kitchen with something like making biscuits. After the biscuits were made, I still played with the leftover dough and scooped flour out of the twenty-five-pound flour bin. While I always liked doing things with her, it was no match for things on the farm.

One afternoon I made my way to the barn lot where the mules were loafing. Since all the menfolk were gone, I decided it would be easy to lock the mules in the barn and catch Jack for a ride.

Everything went as I had planned. After locking Jack in his stall, I let the other mules out of the barn and closed the door behind them. Jack and I had the barn to ourselves. I got his bridle out of the harness room. The bridling went as I had planned too. I stood in the hay manger to be high enough to put the bridle on him. He opened his mouth on cue to accept the bit, then I reached as high as I could to slip the head band over and behind his ears.

The hay manger also gave me the right position to climb onto his back. Pulling on one of the reins, I turned him and he started walking around in the spacious barn. I had a false sense of safety as I rode him in circles. He couldn't run away with me. We were locked in the barn. I practiced my gee and haw commands and even improvised on a right turn with "Gee Jack, I sure do like you." I did giddy up and whoa, but never did kick him into a trot, fearing I would bounce off, but I did encourage him into a fast walk.

Having satisfied myself with a successful mule adventure, I unbridled Jack and turned him back into the barn lot. I don't think anyone ever found out I had ridden Jack; even if they did, no one said anything.

I've been left with the thoughts of what could have happened in my adventures with Jack. My last ride on Jack could have been my last ride on anything in this world.

First Shall Be the Last

I know you think the title should be "The Last Shall Be First," but not in this case. My first hog was my last. As a young boy, summers and school holidays often included a week or two of staying with grandparents on their farm in Henderson County. It was my love of country life that started there, drew me there, or both. I never passed up a chance to go to my grandparents and usually schemed to make it happen when there weren't obvious opportunities. The desire was only diminished slightly when we bought our farm in the summer of 1959. The activity of our ten acres couldn't compete with the excitement of livestock and hundreds of acres at my grandparents. So even that summer, I was still happy to go spend a week there. This summer I would bring home more than chiggers, sunburn, and memories.

My granddad knew of my love for animals. He had been my partner at age eight when I won fifty baby chicks and had no place to raise them in town. He borrowed a neighbor's pony when I was nine so I could ride all week. It wasn't unusual for him to recognize that our ten acres was incomplete without some animals. On the first day of my visit that summer, he pointed to a field of weanling pigs below his house and told me if I could catch one while I was there, I could take it home with me.

Even at that age, I was known to be a fast runner. My dad called me Speedy. The thought of catching a weanling pig seemed like a sure thing. All week long I chased pigs. Several times a day I harassed them. All week long I was unsuccessful. I quickly gave up on trying to run one down in the open field. It wasn't that they were so fast. They were more agile than me. About the time I got close enough to grab one, it would turn and leave me in the dust. I tried sneak attacks, surprise attacks, and luring them close with corn. Nothing worked. I even tried herding a dozen into a corner. I figured that with that many I was certain to get at least one with a sudden grab. That's all I needed-just one! Even then I found little place to grab onto, and they all scattered, leaving me empty handed. It was going to take more than my speed and stealth.

I needed technique. On Sunday afternoon when my parents came to pick me up, I was still empty handed. By then, the pigs would jump up and run away just at the sight of me. I just knew I would be headed home without my gift pig. But my granddad had mercy and told my uncle Arnold to go catch a pig for me. Arnold made quick work of it. He had technique. I noticed he used my cornering approach, but the difference was well timed, fast hands, not fast feet. As Arnold pressed in and made them scatter, he made sure some ran down the fence line on one side. As they ran by, he grabbed one by the hind leg. The pig let out a squeal that made me think his leg was being broken off. How would we ever ride home in our car with a pig squealing like that? Soon the squealing turned into rapid grunting and eventually even that stopped. I still wasn't sure how we would haul it home in our car. All I could imagine was me trying to hold a fifty-pound pig in my lap for a half hour ride home. Fortunately, I didn't have to do that. Arnold got some grass string from a hay bale and hog-tied the pig, and then stuffed it into a burlap sack. They put the "pig in a poke" in the floor board of our car where it rode quietly all the way home. I didn't know it at the time, but granddad told Arnold to catch not just any pig, but a gilt (for city folks, that's a young female pig), so I would be able to use her for raising my own pigs. I was proud to be the owner of my first pig, even if I hadn't been able to catch her by myself.

Our farm had been used to raise hogs at one time. It even had a concrete floor in one of the barn sheds and automatic watering troughs. But by the time we got the farm, the troughs no longer worked and all the fences were down. Locking my new pig in a horse stall was our only option since we hadn't been expecting to bring a pig home. What my pig lacked in proper accommodations was made up for in food. In addition to a daily ration of ears of corn, she also got a slop made from all our table scraps. She was an instant garbage disposal. After a few weeks of living in this 12 X 12 stall, one can imagine what it began to smell like. A pig sty! We quickly went to work on getting the concrete floored pig parlor ready for occupancy. We also put up a new fence on one side of the barn to make a hog lot. With a hole cut into one side of the barn for the pig to move in and out of the pig parlor, an automatic waterer for cool, fresh water on demand, and a large hog lot all to herself, this hog had a good life. This free pig had cost my dad a lot of money. The pig had no respect for her new lot and quickly made it her own by using her snout to root up nearly every square inch of the lot. On a visit a few weeks later, my granddad told me I could stop the rooting by putting rings in her nose. That never happened. I would not be surprised if the damage she did to that lot is still there.

About the only hardship she had were the days I had friends from town visit. I would bet them they couldn't ride her for ten seconds. After I pitched a few ears of corn out for her, getting on for a ride wasn't so hard. Staying there was the problem. First, the back of a hog is rounded from front to back. In front of where one would sit it slopes down toward the head. In the back it slopes down toward the tail. Unless one stays perfectly balanced in the middle, they will slip off either the front or the back. The second problem makes the first even worse. There is nothing to hold onto: no mane, no hair, no horns, nothing. After she started to run, stop, or change directions, one is on the ground looking up at her. It was a sure bet! The bigger the friend, the shorter the ride because she would take off quicker to get him off her back.

In addition to providing city slicker entertainment, she was also the object of my first 4-H project. At the end of the fall semester, I wrote a report on all I had done to raise this hog. My report got me a blue ribbon at the district contest. I'm convinced my description of the arrangements for clean water and fresh air were what gave me the blue ribbon. It sure wasn't the economic profitability of the project.

Around Thanksgiving I was with my granddad again and he wanted to know if I was making any arrangements to have my hog bred. I hadn't even thought about it. I had hardly paid any attention to whether my hog was a male or female. I knew some neighbors who had hogs, but had no idea how to arrange to have her bred. When I mentioned it to my dad, he didn't seem very interested in the idea. A few weeks later he proposed buying the hog from me and having her killed for the family to eat. With the bad smells and all that she had torn up in the hog lot, I had no sentimental attachment to this hog. She didn't even have a name. She was just referred to as "the hog." Since I knew nothing about getting her bred and raising little pigs, the thought of making her my first and last hog sounded appealing. A twenty-dollar bill changed hands, and I was out of the hog business.

A Goat or a Pony?

Children save their money to buy interesting things. I don't know what today's children are saving their money to buy, but it is probably electronic. In the 1950s I was saving my money to buy a pony. The pages of our Sears-Roebuck catalog were folded back to the horse supplies section. There I could pore over the descriptions of the many saddles and bridles offered to decide which I would buy, if I had the money and if I had a pony to put it on. According to Mr. Sears, I would need $150 to buy a pony, saddle, and bridle.

In 1928 when my father, Archie, and his sister, Evelyn, were six and seven years old, they were saving their money, but it wasn't to buy a pony. They wanted a goat! They earned money by selling the eggs they found in hidden nests in the barns. After a couple of years of saving their egg money, they had enough to buy their goat. Their dad would often have a few goats, not for pets, but to slaughter and barbecue on special occasions. Archie and Evelyn's goat was all for fun. They named him Billy. It wasn't a very original name, but children aren't always interested in originality when it comes to naming pets. Unlike their dad's goats, Billy was able to roam free, just like a family dog or cat. Sometimes Billy would be missing and someone would ask, "Where's Billy?" Archie and Evelyn would start a search that usually ended with Billy being found with his head stuck through the hole in a woven wire fence. There had been room for him to reach his head through the square space in the woven wire as he reached for some grass on the other side of the fence, but when he was finished eating and pulled his head back, his horns would catch on the wire and he was stuck. His only escape was to stand there and wait for someone to free him by tilting his head back to let the tips of his horns pass through the wire first.

The children wanted Billy for more than just having an animal they could pet. They wanted free rides. Their grandfather made a leather harness for Billy and a cart for him to pull. At first Pappy would put Archie and Evelyn into the cart and lead Billy pulling the cart. With a little training and a few exciting wild rides, Billy soon learned to be driven by the children and they were free to ride in their goat powered cart wherever and whenever they wanted.

Billy could do no wrong, or so the children thought. As Billy aged, he developed more bad habits. The most common, and the one most goats have a reputation for, was to butt things with his head, especially the rear ends of people who weren't looking. Since Billy was bigger than the children, butting them usually meant they were knocked to the ground and tears would flow. Billy could also be provoked to butt an unsuspecting person when someone in front of them teased Billy by shaking their finger at the goat. It was a prank that quickly fell out of favor with the family. More than one screen had to be replaced because of Billy running through the screen door to get at someone. In spite of his bad habits, Billy was saved from the barbecue pit. Who would barbecue a family member just because he misbehaved?

Billy was still around when Archie got his first car. It was a 1947 black Chevrolet convertible, and Archie was so proud of it, that he used every free minute to keep it clean and polished. It shined so much, Billy saw his reflection in the door, thought it was another goat, and gave it a butt. Archie was so mad he threatened Billy with the barbecue pit.

Head-butting wasn't Billy's only natural behavior that caused trouble for Archie. One day Archie was changing a flat tire on the car. When he took each of the six lug nuts from the rim, he put it in the upturned hub cap on the ground nearby. Archie put the spare tire onto the hub and reached for the lug nuts, but they were all gone. Billy had eaten all of them. Car owners keep spare tires, but not spare lug nuts. Archie took one lug nut off of each of the other tires to temporarily replace enough of the ones Billy had eaten until he could go to town and buy new lug nuts.

The family kept Billy until he died, but one incident really put the taste of barbecued goat into Archie's mouth. It was another incident involving Billy and Archie's beloved convertible. Archie was in the house and was surprised to look out the window and see Billy standing on the hood of the convertible. For Billy, the car was just a shiny mountain to climb. If Billy couldn't butt it or eat it, he would climb it. A slow, calm approach would have been best, but Archie wasn't in the mood for either. He tore out of the house to chase Billy off the car. Billy did what most of us would do if we were being chased by a mad man. He moved to higher ground. Up onto the cloth convertible top he scampered. Billy was out of reach and king of the mountain. He didn't stay king very long because the combination of his weight and pointed hooves sent his legs ripping through the fabric top. First one leg fell through, and then trying to regain his balance, he put more weight on another leg until it fell through. Finally, all four hooves punctured the fabric, and Billy's stomach was being held up by the cloth top as if in a sling with his legs flailing inside the car. In one last convulsion, Billy broke completely through the shredded top and was left standing on the back seat. Archie opened the car door, and Billy stepped out like he had just arrived home from a drive and his chauffeur was there to assist him. If Archie wasn't thinking of killing Billy at that moment, I'll bet he was at least wishing he had saved his money to buy a pony. Now he would have to use his money to buy a new top for his convertible.

Baldy

Today's little boys dream of having four wheelers, dune buggies, and jet skis. They want the feel of speed and acceleration. They want the smell of gasoline and hot engines. In the fifties little boys wanted a pony. They wanted the feel of slick hair and a velvet nose. They wanted the smell of new leather and grassy breath.

In elementary school I would daydream about organizing pony races on my Uncle Ronnie's farm. He had a field full of ponies that were used to pull carts of coal from the western Kentucky mines. I didn't know who I would race against, but I knew I would ride the chestnut stallion, Dan. He was blind in his left eye from a mining accident and always had to be reined for a right turn, but he was still the fastest in the field. I had ridden him on a summer visit, and when I was on him, we seemed to effortlessly sail across the pasture. It seemed effortless to me because Dan was doing the running. I could go pretty fast on my bicycle, but like riding a stick horse, I had to do all the work. On Dan all I had to do was hold on, feel the wind in my face, look ahead for a right turn, and duck my head when we went through the low stall door. This was nothing like being at the pony rides at city park, where kids just sat there riding around in circles atop their sleepy steed, whose bridle was tied to one of the revolving poles. There was no wind in their faces at this speed, nor any freedom to explore the fields. It was just a continuous right turn and one step above the fiberglass pony at the front door of the grocery store. I always climbed into its saddle and imagined I was on a live pony. I whipped his neck with the reins but we just sat there motionless. Even when Mom splurged and dropped a nickel into the control panel and the pony gave me a minute of its rocking, simulated gallop, I was left unsatisfied as it ran out of time and slowly ground to a lifeless halt.

My dream was fueled each fall at Wyndall's Grocery where I could put my name on a piece of paper, drop it into a wire mesh barrel and imagine how wonderful it would be if I was the lucky winner of this year's pony, saddle, and bridle. During the weeks of build up to the grand prize drawing, Wyndall Smith had the bridle and saddle on display by the front door near the mechanical pony. On the weekends, even the live pony was on display in the parking lot. I never won, but I kept stuffing my name into the barrel and dreaming I would win. It looked like I would have to get my pony the old-fashioned way. I would save my allowance and buy one.

My allowance was fifty cents a week. I would get it on Saturdays, but rather than actually being given two quarters, Dad suggested I keep a ledger. Every Saturday I would add fifty cents to the balance I recorded in the General Electric calendar diary he had brought home from his work. When I wanted money, I would get it from Dad, the banker, and subtract it from the diary balance. In addition to keeping up with my account balance, it gave me a record of what I spent my money for. This was big time accounting for a nine-year-old. It was probably good that I couldn't multiply or divide yet because I would have figured it would take more than six years to save enough to buy a pony. I knew exactly the cost of a pony, saddle, and bridle because they were pictured and priced in the Sears-Roebuck catalog for $150. After a while, I had no trouble finding the page where they were listed along with horse saddles and harnesses. The catalog opened automatically to the page, which was dog eared from repeated use.

The next summer my pony dreams were fueled by my regular summer visit to my grandparents for a week on their farm. Granddad had borrowed a neighbor's pony for me to ride that week. It was a week of heaven for me. I rode that pony several times a day every day. For the week, my dreams were real. The only part of the dream for a pony of my own that I hadn't figured out was where I would keep a pony when I got one. It didn't seem to matter to me that we lived in town in a subdivision that wasn't going to allow a pony in the backyard. Who can be bothered by details when you're having big dreams?

I listened with interest when my parents started talking about buying a farm and moving to the country. I got all excited when we actually went to look at one and disappointed when we drove away with Dad saying this wasn't the right one. But they soon found the right one, and we moved eighteen miles from town to ten acres with a house, lake, and most importantly, a barn for a pony. We were hardly unpacked before I started asking regularly when I could get a pony. Fortunately, I didn't have to wait but a few weeks before we took a Sunday afternoon drive to an unannounced destination. As soon as we pulled into a driveway and I saw a pony in the front yard, the smiles on my parent's faces told me the purpose of the trip. The pony was a dark brown Shetland with a white face from between his ears down to his muzzle. I would learn later this was called bald faced, which is why he was named Baldy. He was already saddled and ready to ride. I didn't need any encouragement to take Baldy for a test ride around the yard. I wasn't a great judge of horse flesh, but I knew the best thing about Baldy was he could be mine. The moment of truth came when Dad asked the owner how much he wanted for the pony. There was also a cart and harness, but I knew there was no way I would have enough money for those too. The man said he would take $150 for the pony, saddle, and bridle. He must have been reading the Sears-Roebuck catalog too! Dad asked me how much money I had and I sheepishly said, "Only $57.60." I couldn't believe my ears when Dad said, "We will take him and I'll pay the difference."

Baldy wasn't the most handsome pony I had ever seen, but he was mine. He wouldn't neck rein like some ponies I had ridden, but he was mine. He wasn't as fast as Dan, but he was mine.

That winter Dad got a mule harness from my grandfather and cut it down to size to fit Baldy. Dad and I would cut locust trees with a crosscut saw for fence posts and we would have Baldy pull them out of the woods. Baldy pulled hard and wouldn't stop until we were where we wanted the log to be.

In the summer Dad hitched Baldy to a three-point cultivator to plow the tobacco. The summer before, Dad had borrowed a big black Percheron draft horse from our neighbor to plow the tobacco. The horse had plenty of power for the plow. Unfortunately, it hadn't been worked for several years and it wanted to trot down the rows. Dad could hardly keep up with him. It was a funny sight. It was Dad who had to stop from his jogging down the row for a rest at the end. Even though Baldy had to stop for a brief rest at the end of each row, his slow pace got the job done.

I soon got the chance to drive Baldy myself. There was more to pay attention to than I realized. I probably plowed up half of the first row because I couldn't control everything at once. As if it was enough to just hold a rein in each hand, I also had to have a plow handle in each hand too. Baldy probably knew more about walking straight down the row than I did, so fortunately he didn't require much reining. Directing the plow seemed to work backwards for me. When it drifted too close to the row, I would pull the handles away from the row, but the plow moved even closer to the row. I moved the handles even further away and then the plow was on top of the row and I was plowing up the tobacco plants. I pulled on the handles to drag the plow off the row, but it became a tug of war between me, the plow, and Baldy and they kept winning. Dad walked behind me and gave me some pointers, and at the end of the row he gave me some pointers on moving the plow handles in the opposite direction I wanted the plow to go. He plowed the next row as I walked behind him watching him effortlessly control Baldy and the plow. He handed the reins back to me and said, "Here, you need the experience." I got to do a lot of things like that, because I seemed to always need the experience. Even though I still wasn't very good at plowing a straight row, I at least could avoid plowing up the tobacco plants.

Even with stopping at the end of each row for a rest, plowing the half acre of tobacco was really too big of a job for Baldy, so his pulling was limited to a sled Dad made for moving things around on the farm. Cultivating the tobacco progressed from the Belgian horse, to Baldy, then a rotary tiller, and finally a Super A Farmall tractor that was made for the job and didn't have to rest at the end of each row.

My time with Baldy was spent on his broad back and not walking behind him. He could walk lazily along without my holding the reins or we could fly up and down the hills at a gallop. It was even better than my dreams.

Baldy wasn't a show pony, nor did I have any training in proper riding technique, but that didn't keep me from entering him in the pony class at the local horse show one summer. Baldy and I spent weeks practicing together. We went through each gait. First, I had him do a walk and I focused on pulling back on the reins to keep his head up instead of him walking along with his head down like a tired plow horse. When he trotted, I had more to learn than Baldy. I had to learn to post in the saddle at the same rhythm as his trot so I didn't bounce up and down in the saddle. Baldy and I had spent so much time galloping through the fields; he wanted to break out of a canter and into a gallop.

We didn't have any electric clippers, so I used scissors to trim his mane. Baldy's mane wasn't the long thin type that draped over his neck. It was thick and bushy and stood straight up like the mane of a wild Mongolian horse. It looked best when it was closely cropped like a burr haircut. Using scissor was a very slow process. I had seen our neighbor, Joe Tom Taylor, use his pocket knife to cut a hand full of hair from his plow horse Rosie's mane or tail, so I thought I would try that. It was faster and worked pretty well until I lost control of the knife at the end of a cut and sliced a gash across the top of Baldy's neck. Baldy hardly flinched, but I was horrified at what I had done as I saw the thick red blood ooze from his neck and through his mane. That ended the mane trimming for the day and I was back to the scissors for the rest of the trimming. I would reserve the knife for the end of Baldy's tail.

The day of the show I cleaned and polished the saddle and bridle with saddle soap until it shined. I put brown shoe polish on a spot or two where the saddle had been scratched by getting too close to a tree. Baldy got a complete shampoo and rinse from the garden hose which he liked except for getting sprayed around his ears.

It was only a mile and a half from our house to Pleasant Ridge Elementary School, where the horse show would be held on the ball diamond. I rode Baldy along the roadside of US Highway 231 to get to the school by seven o'clock. The sun didn't go down until late during the summer, so it was still daylight.

The under forty-eight-inch pony class was the first class of the show; I didn't have to wait long for our performance. All the horses and ponies rode in the grand entry and lined up for the playing of the national anthem from a record over the loud speakers. I didn't have a cowboy hat, but all the Western riders removed their hats and put them over their chests while the anthem played. I just put my hand over my heart like I did when I said the pledge of allegiance at school. The flag horse led us out of the ring and we were ready for the first class. I was really nervous and didn't feel any better as we lined up and I looked around at the competition. The other riders were in fancy riding outfits, especially the girls, and they sat so erect in their saddles. I was wearing my jeans and a T-shirt but did my best to imitate them by sitting erect in my saddle too. Baldy was the only brown pony in the ring. All of the others were flashy pintos, sparkling whites, or shiny blacks with flowing manes. While I hoped the judge would place Baldy and me among the top five so I would get a ribbon and some prize money, I wasn't really surprised to only get a participation ribbon. I tied the ribbon to Baldy's bridle and spent the rest of the evening showing him off to my friends. I let them take turns getting on Baldy and being led around the playground. I couldn't trust these unknown riders in a crowded environment to ride by themselves, so all they got was a ride with me leading them around like the pony rides at city park. This was enough for them, since all they had were dreams of one day being like me and having a pony of their own.

The horse show didn't end until eleven o'clock and riding home in the dark was going to be a lot different than the afternoon ride to the show. All of a sudden, I wished we had a trailer or even a pickup truck to haul Baldy back home. I probably could have gotten a ride home from someone with extra room in their trailer, but I didn't know anyone and was too shy and independent to ask for a favor from a stranger. It would only be a twenty-minute ride, but I had never ridden Baldy at night and being on the roadside didn't make me feel any better. Mom and Dad drove behind me in our '57 Chevy which helped light my path and protect me from traffic coming up behind.

Baldy and I started off at a walk, but I knew it would take a long time to get home at that pace. For the first half mile we trotted along the county road where the traffic was slow and light, but when we got to US Highway 231, there were a lot more cars and they were going fast. Even worse was meeting a semi-truck which made me feel like I was about to get blown off Baldy's back. I just hoped the noise of the big trucks didn't scare him and make him run into the ditch, or worse, into the road. This was the part of the ride I had been dreading. I moved Baldy over into the grass of the roadway and kicked him into a gallop. Moving faster through the damp chill of the night air made me wish I was wearing a jacket instead of just a T-shirt. An occasional bug would hit me in the face. My glasses served as a good windshield for my eyes. The bright headlights from oncoming cars would blind me momentarily as they met us, and I hoped Baldy could see the roadside better than I could. He never faltered and kept running straight ahead with that "going to the barn" gallop. Reaching our driveway was a pleasant sight, and I reined Baldy into a walk. As we walked up the drive and could finally relax, I realized how tense I had been. Baldy seemed relieved too, as we made our way toward the barn. He gave a snort to clear his nostrils as if to say "it's good to get home." Back at the barn he crunched on the extra ears of corn I gave to him, as a reward for the hard work. I uncinched his saddle and, when I removed it and the saddle pad, steam rose from his back. It had that familiar smell of sweaty pony hair. It wasn't a bad smell, but one that I had grown to like. I breathed it in deeply as I brushed down the wet hair. The smell was just another reminder that this wasn't a dream. I really did have a pony of my own. Any dreams I had as I slept that night couldn't be as good as the real thing.

What Was I Thinking?

Some experiences belong in the category of "What Was I Thinking?" Most of them come down to the fact that there wasn't enough thinking going on. When animals are involved, it usually means we hadn't bothered to see things from the animal's perspective or didn't know enough about animal behavior to anticipate the results of our actions on them.

There were several "What Was I Thinking?" experiences with my first Holstein heifer. The first unexplored question was why I got her in the first place. Dad and I were already partners on several Hereford cows and a bull. We made good money selling their calves every fall. I had learned in my tenth-grade vocational agriculture class about this new breeding process called artificial insemination. I convinced Dad that we should sell our bull and breed all of our cows artificially. We wouldn't have the expense of keeping a bull and could breed our cows to the best bulls available. What was I thinking? I knew little about the importance of timing in breeding or about heat detection.

About the same time, I heard about several people who had Holstein dairy cows and would raise several calves a year from one cow by buying three day-old calves from dairies and letting three calves nurse one cow. After a few months, I would sell the calves for veal and put three more calves on the cow. What was I thinking? I had no dairy cow experience, except milking our Jersey cow when I was twelve. My only experience of buying two baby calves to raise on a bottle didn't turn out so well either, when both died of pneumonia and scours. What was I thinking?

My vo-ag teacher, R.C. Johnson, and his brother, Larkin, were in the dairy business and I proposed to him that we trade my pony for one of his Holstein heifers. I thought it was a great deal for both of us. I had outgrown the pony, he had two young daughters, and there were lots of fine Holstein heifers on their farm. The values of the pony and a heifer were about the same, so we were both happy with an even trade. What was I thinking?

Dad and I drove the ten miles to Hartford, where the Johnsons lived, to deliver the pony and pick up our new heifer. I was proud of the new cattle racks on our truck. I had made them as a project in the vo-ag shop and got a blue ribbon on them at the county fair.

All of the loading and unloading went fine. But, on the trip back home, we noticed the heifer was very restless. She kept turning around and looking for a way out. Fortunately, the racks were plenty high and she couldn't escape. Her being nervous made us nervous. When we pulled into the pasture to let her out of the truck, she was still moving around and looking for a way out.

Dad said that with all of her energy and excitement, she might need something to keep her from jumping a fence. I had seen pictures of cows wearing a forked stick around their neck, like a yoke with the single end of the stick hanging down between their front legs. It allowed them to put their head down to eat and drink, but was a barrier if they tried to go through a fence. Dad said when he was still on his parents' farm, they had a new mule that was wild; they put a log chain around its neck and tied the chain to a fence post for it to pull around for a few days until it calmed down. He suggested we do that to this wild Holstein heifer. This time it's "what was he thinking."

With the log chain around her neck and the other end of the chain tied to a fence post laying on the ground by the truck, we opened the truck rack gate and let her jump out. At first, it looked like everything would work out as we planned. Feeling the weight from the chain and post, she moved away at a fast walk to put some distance between her and the truck. That's when we began thinking about what she was thinking about. "Something is following me. What is it? Maybe I can get away by out running it."

And run she did. She made a wide circle around the pasture at a speed I had never seen a cow move. Not having any luck losing this thing behind her, she decided to jump a barbed wire fence into the tobacco field with hopes of leaving her pursuer in the pasture field behind. Even pulling the weight of the fence post, she cleared the fence like an Olympic athlete. It was the post that did all of the damage as it ripped every strand of wire to make a new opening. She also made a new opening down the middle of the tobacco patch as the post crushed the plants that she didn't trample.

Running in pursuit through the new gap in the fence, we tried to catch up with her. What were we thinking? She was flying like the wind and motivated by sheer terror of the relentless post following her. Besides, what would we have done if we caught up with her? Before we could clear the tobacco patch, we heard another crash at the other end of the field and knew from the sound that she had gone over and through the fence at the other end of the field.

We were relieved when we got to the fence to discover, not the heifer, but at least the chain and fence post stuck in the fence. Fortunately, the piece of rope tying the chain together around her neck had given way to the pressure of the post getting caught a second time in the fence. We looked down the hill in time to see the heifer sail effortlessly over our property line fence and disappear into our neighbor's one-hundred-acre corn field. There wasn't another fence for miles. Who knew how far or where she would run before she stopped? We surveyed the fence and crop damage as we walked back to the truck. I'm sure we also wondered, what were we thinking.

Later, I went back to the spot we had seen her first sail over the fence and began tracking her through the corn field, down an eight-foot creek bank, along the creek, and then up the bank on the other side into another neighbor's hay field. There was no heifer in sight. It was beginning to get too dark to follow her tracks in the hay field; so, I walked the mile back home.

The next day we got a call from a farmer who lived two or three miles away. He had been calling all over the neighborhood trying to find out if anyone was missing the Holstein heifer that had jumped into his barn lot. He said he didn't have a clue about where to start calling, because he knew no one in the neighborhood had any dairy cows. We admitted she was ours and told him we would come to get her.

She seemed a lot calmer when we loaded and hauled her this time. It was probably a mixture of being worn out from her escapade the day before and the relief of not having that post following her. This time we just pulled the truck into the pasture, opened the rack gate, and let her hop out. She walked a few feet away, put her head down, and started grazing. After a few bites, she raised her head, chewed on a mouthful of grass, and gave us a look of "what were you thinking?"

To Market, To Market

Getting cattle from the farm or ranch to market has seldom been simple. Cattle drives had their stampedes, strays, and rustlers. It was expected to lose several along the way to injury, death, theft, or escape. The introduction of trains, big trucks, and cattle trailers was an improvement, but they also had their problems. Cows get excited when you take them out of the wide-open spaces of the pastures and crowd them into a small space in a corral or barn. The adventure starts even before they are loaded onto a truck or trailer. An especially wild cow may not go to market as planned because it breaks the planks from a pen at the barn, jumps the barn lot fence, and sprints with its tail and head high in the air back to the pasture. It's a disappointing and frustrating sight, but a vivid reminder of where the term 'high tailing" it originated.

In the summer of 1966 before I went off to college, Dad and I decided to begin selling our cattle to help pay for my college expenses. Since all of the cows were bred, the first to go was our nearly two thousand-pound horned Hereford bull. Unlike today, no one had a cattle trailer in those days. Being low to the ground, they make loading easy and can be pulled by a farmer's pickup truck. Before trailers, cattle were hauled in pickup trucks, if you only had a few to haul. For large loads we called Carl Hoover, whose big truck was the neighborhood's way to get livestock to market.

A couple of years earlier during a school project, I had made a set of wooden racks to go onto the back of our pickup truck. The light weight pine boards made them easy to put on and take off the truck when we weren't hauling cows. They had already been put to good use hauling steers to shows or freezer beef to slaughter. That's how we were going to haul the bull to market.

Before daylight one August morning, our neighbor, Sid Freels, came over to help me load and haul the bull. The day started off hot, but the bull walked calmly up the loading chute which was level with the truck bed. As he stepped onto the truck, we knew we had a heavy load. The truck springs took on the bull's weight and the truck bed lowered nearly a foot. The bull weighed nearly as much as the truck. As big as he was, there was still enough room for him to turn around in the truck and he made it rock in every direction when he moved. He raised his head above the racks but made no move to jump out. Sid had seen enough animals escape over tall walls and fences to know keeping their head down would prevent an escape. He suggested we tie the bull's head to the truck which would not only keep his head down but keep him from turning around. I knew we couldn't tie him to the light weight boards of the racks and none of our old ropes would hold this big bull, so Sid suggested we use a log chain. He wrapped the chain around the bull's ten-inch horns and the other end of the chain was slipped through the racks and down to the frame of the truck. This bull was not going to get away from this truck.

As I drove easily out of the barn lot, the bull, Sid, and I were all calm and relaxed. I did notice that with all the weight on the back of the truck, its front pointed up a bit and the steering wasn't as positive as usual. I would have to stay alert to keep this truck between the ditches. My biggest concern was when the bull shifted his weight from one foot to the other and the truck also darted in the same direction. Most of the fifteen-mile trip was along country back roads. While the road was paved, there were a lot of curves and hills to negotiate. The bull settled down and shifted his weight less often as we picked up speed to forty miles an hour. That's as fast as I wanted to go and feel confident of maintaining control of the truck. Even then, to keep the truck on the road I had to slow down when we came to curves. Because of the heavy load, we would lose speed going up the hills and would have to brake to keep from going too fast down the hill. Sid and I realized we were over loaded, so I drove sitting on the edge of my seat. One instant of not responding to a change in the truck's speed or direction and its momentum would have sent truck, men, and bull off the road and into a ditch or field.

All went well until we were going up a very long, steep hill. Even with the truck in first gear and the accelerator on the floor, the truck kept getting slower and slower. By the time we were nearly to the top of the hill, we were only going ten miles an hour, but I judged we would make it over the top. That's when the bull started getting restless. When we were cruising along at forty miles an hour, he kept his eyes on the road ahead and had no thoughts of jumping out, but as we slowed, he started looking from side to side like he was trying to spot a place to land. He shifted his weight first right, then left and back again. The truck swayed with each move. We were thankful for that log chain around his horns. But with a big lurch to the left, we heard a crash and turned to see the bull toss his head into those light-weight pine boards of the side rack. With splintering boards flying onto the road, he jumped through the hole he had made with his head, making it larger as he jumped and landed in the road.

Fortunately, there was no traffic on this little country road, because the truck had one lane blocked and the bull blocked the other. He tried to get away from the truck, but we were glad to see the log chain around his horns was still doing its job and he couldn't move more than a couple of feet away from the truck. At least the bull couldn't run away. We had enough problems on our hands without a runaway bull too. We got out of the truck to evaluate our situation. Sid bragged on our decision to use the log chain. The relief didn't last long. Being closely tied to the truck had made the bull mad. This truck scared him and he wanted to get away from it. The log chain now shifted from friend to enemy. First the bull butted the truck fender. I crawled over the racks and into the truck bed and yelled at the bull to get him to back away from the truck and stop butting it. It worked for a moment. With an angry snort he lunged up at me with his mouth wide open and bawling like he wanted to eat me alive, but the chain caught him short and he slid back to the pavement. I fully expected him to lunge at me again, so I backed away hoping he would calm down, but he didn't. He lowered his head to the pavement and put his head under the truck. Using the muscles of his massive shoulders and neck he raised his head back and forth. The truck rocked so much, I had to hold onto the racks to keep from falling. I thought he was going to turn the truck over. He was as mad at me and the truck as I had ever seen a bull, before or since. Sid yelled, "We have to unchain him or he is going to tear up the truck!" Fortunately, the bull gave up on trying to turn the truck over and just stood their staring at me with his wild eyes and his sides heaving, trying to catch his breath from the exertion and excitement.

Cows don't sweat, but if they did, he would have worked up a lather by now. His mouth hung wide open as he tried to get some fresh air and cool off. It's easy for a cow to get super-heated and can lead to their death. Through fear, exertion, or the hot August air, I don't know which, I was beginning to sweat now. How could we get close enough to the bull to unhook the chain without him using his horns on us? While Sid was circling behind the bull and thinking of what to do next, the bull stuck his head back up to the hole in the racks where I was standing. I thought he may be coming to get me again. He was just stretching his neck and head out to investigate. I used the opportunity to quickly grab the hook on the chain. By the time the bull jumped back, I had unhooked the chain and it slid from around his horns falling to the ground with a rattle. It took him a moment to realize he was free and he just stood there looking at me. It was as if he were deciding which would bring him more pleasure, using his new freedom to high tail it from this crash scene or jump back into the truck and punish me. Fortunately, he chose to high tail it. He turned around, gave Sid a wanting look, jumped the ditch, and trotted off across an open field.

For the moment, I was relieved to see him go. There were probably a thousand acres of open land around us with no fences or buildings in sight. How would we catch him? How would we load him and into what? How would we get him to market? At the moment I didn't have answers to any of those questions, but was feeling relieved. As Sid and I watched the bull trot further away, over a hill, and out of sight, we quickly discussed our options and decided I should drive the truck to find a house where I could call someone for help. He would follow the bull and keep track of it.

Carl Hoover to the rescue! His wife answered when I called and she said Carl was in his shop getting the tractor and equipment ready to bale some hay for a neighbor, but she would send him to help us first. It would take him about a half hour to get there, so I went back to look for Sid and the bull. I found them a quarter mile off the road. It was quite a sight. They were both standing in the shade of a big maple tree. The bull had his head down and had caught his breath as he calmly swatted a horse fly with his tail. Sid was nearby with his straw hat in one hand and a small tree limb in the other. I didn't know and didn't ask if the limb was for protection or to fan the bull with a cooling breeze.

We slowly herded the bull at his own pace for three quarters of a mile back to the house where I had called for Carl Hoover. Even though there were no fences along the way, we were able to get the bull into their stable about the time Carl Hoover drove up in his big truck. There was no loading chute in the stable, so how would we get the bull up three feet into the truck? Having hauled a lot of livestock to market under less than ideal conditions, Carl knew all the tricks. He backed his truck up to the stable and it filled the alley opening. He made stair steps with three bales of hay stacked behind the truck. We all got behind the bull and herded him down the alley toward the truck. When he got to the strange looking makeshift hay bale steps, he hesitated just a moment evaluating the situation. Should he eat the hay, climb it into the truck, or turn around in retreat? Carl used another of his cattle loading tricks and didn't give the bull long to consider his options. Like a cowpoke, he poked the bull on the rump with his cane and the bull clambered up into the truck. Carl slammed the truck tail gate closed, and the bull was now where he should have been at the beginning of the morning. Our ordeal had ended. Carl drove confidently and happily away to market with our bull, as if it was just another ordinary day. It had been no ordinary trip to the market for Sid and me, but then it seldom is.

Sunday Dinner on the Grounds

When I was a boy, it was the tradition for Sunday dinner to be the biggest meal of the week. When I say dinner, I mean the noon meal, as in breakfast, dinner, supper. Mom would devote much of her day off from work to prepare dishes that couldn't be done in the little time available between arriving home and the time the starving family expected supper to be on the table. Even though quickly prepared TV dinners were available, they were frowned upon as being low class, probably because they were low class. Throwing something into the microwave, zapping it in a few seconds, and calling it dinner was only something of dreams.

Right after breakfast and before we headed off for church, Mom was already at work preparing Sunday dinner. While we were away, the salads could chill and beans could simmer, but the tool that really saved the day was the pressure cooker. With the combination of pressure and steam heat it cut in half the time needed to cook a roast or chicken. The pressurized moist heat would turn tough meat into tender meat a lot quicker than slow roasting in the oven. With hungry, impatient kids waiting, a slow roast was not the path to a happy family dinner on Sunday. As the meat cooked, steam escaped through a small port on the pressure cooker top. The temperature and pressure were controlled by a weight that was placed over the port. Each side of the weight had a different sized opening so there was a choice of cooking pressures. As pressure from the steam built up inside the cooker, it would raise the weight off the port slightly and allow some steam to escape. When the cooker really got going, the weight would jiggle up and down and make a sound heard throughout the house that everyone soon associated with good things happening in the kitchen. When the cooking was done, the stove burner was turned off and the weight removed to allow the steam to escape like steam coming from a train whistle. With the pressure reduced, the top of the cooker could then be safely removed.

This cooking process still took several hours, but Mom could put the pressure cooker onto the stove and let it cook while we were away at church. The meat would be fully cooked when we returned. All that was left in preparing dinner was mashing the potatoes while the Brown and Serve rolls baked in the oven. A roasting hen would come out of the pressure cooker with its meat falling off the bone because it was so tender.

One Sunday, Mom had started the hen cooking before we left for church. When we returned, we were greeted by the smell of freshly cooked chicken, even before we opened the back door and came into the house. The nearer we got to the kitchen, the hungrier we were made by the smell of the chicken. Stepping into the kitchen, we discovered the chicken had already been served up and I do mean up. While we were away, the pressure cooker exploded, sending its weight and top to the ceiling. There was a mark on the ceiling where the top's port gouged a hole on impact. There was chicken everywhere. It was on the ceiling and light fixtures, all over the stove top, along the top and side of the refrigerator, on the walls, on the window curtains, and on the floor. Months later we were still finding pieces of chicken in unsuspecting places.

Mom was wondering how we would ever get the mess cleaned up. My brother and I were wondering what we would eat for dinner. Dad was evaluating the ceiling to determine if a repair would be necessary. Mom called us all to the task of helping with the cleanup with a warning of not to eat any chicken from the floor. With a chicken wing in one hand and a drum stick in the other, I made my way to the back door, where I let in Janie, our Beagle hound. She had a field day cleaning all the chicken parts from the floor which was off limits to the humans.

Pressure cookers and big Sunday dinners are out of fashion, but I'll bet the current residents of our house wonder about that mark on the ceiling above the stove.

Murder in Chicken City

My daughter, Whitney, was nine when she got her first Easter chicks. The twenty-five newly hatched, fuzzy chicks arrived at our post office in a cardboard box with air ventilation holes. Our farm already had cows, horses, ducks, peacocks, farm dogs, and barn cats for Whitney to enjoy, but these chicks would be hers to care for. They would be her project for lessons on responsibility, entrepreneurship, and murder investigation.

For the first few weeks the chicks lived in the feed room in a four-foot square wooden box under the warmth of a heat lamp until their fuzz turned into protective feathers. This gave us time to prepare their pen and chicken house.

The 20'x 30' dimensions of the chicken pen were determined by how much perimeter could be encircled by the 100' roll of new chicken wire. At 6' high, it would contain the chickens unless something spooked them and the added adrenaline helped them fly high enough to clear the fence.

A red metal storage building that had been left behind by a former renter of the cottage would serve nicely as the new chicken house. Whitney's Easter chicks weren't ready to start laying eggs yet, but before they were ready a row of nests would be added along one side.

While I added the roosting poles inside the chicken house, Whitney put the finishing touches on a hand-lettered sign that would hang over the chicken pen gate. The name for the chicken house and pen had been settled on around the dinner table as plans for the construction developed. Using a three-foot-long plank that had been painted the day before with some left-over chocolate brown paint, Whitney dipped her paint brush into the bucket of white paint and added six-inch letters one by one. Her sign lettering may not have looked professional, but for a nine-year-old it sure looked authentic. When she was done, we nailed the sign over the gate for all to know they were entering CHICKEN CITY!

After three weeks of living in a coop, the chickens had grown to the point of needing more space for eating, roosting, scratching, nesting, running around, and all the other things chickens do. With much excitement on the chicken's part, we moved them to Chicken City.

For the next three years, Chicken City was headquarters for Whitney's poultry operations. There was a family lottery to see who could guess the date the first egg would appear. The eggs started coming in September. At first there was just one or two little pullet eggs a day and some days there weren't any at all. A tan wicker basket with a hoop handle was kept in the mud room by the back door for trips to gather eggs. In the winter when egg production was low, Whitney could gather the eggs and stick them into her big coat pockets. This worked really well except when she went to another chore and forgot about the eggs. We found there's no easy way to remove scrambled eggs from a coat pocket. By March, eggs were coming faster than we could eat them. It was time to unload the surplus on neighbors, family, and friends. "Eggs by Whitney" was born and at a dollar a dozen a new entrepreneur had been unleashed on the world

When Whitney was twelve, the farm was sold and we moved an hour away. The chickens were put into a coop and trucked to the new farm. Other than the chickens, the only part of Chicken City that made the move was the sign over the gate. The new farm already had a large fancy chicken house with roost poles, nests, and an adjoining brooder room for starting new chicks. A larger pen was built, and the following spring, another twenty-five chicks came in the mail. Chicken City had grown from a small village to a metropolis! The larger flock meant more eggs, more customers, and more income. Whitney faithfully made afternoon trips to Chicken City to gather eggs, check on feed and water, and take time to talk to her hens. Her slow movements and calm voice made the hens comfortable enough to come close, eat out of her hand, and even be picked up and held. Some were singled out because of their personality or markings and given appropriate names: Lacey, Red, Blackie, Bashful, Joseph, Eleanor, Jonah. Even if some didn't get a name, their dependency on the daily visit created a bond between fowl and friend. Such a bond only made the coming events more difficult.

The first event came without warning. One afternoon during the spring, when egg production was at its peak, Whitney made her daily egg gathering visit to Chicken City. Stepping through the door and into the chicken house, she was startled by the sight of a dead chicken under the roost poles. The death of a hen from time to time was to be expected, but it always makes ones' heart drop when one is discovered. Moving closer for a better look was even more startling. Its head was missing! This wasn't death by natural causes. The scene smelled of foul play (or even fowl play!). Other than having its head missing, the hen appeared to have just laid down and died. There weren't other visible injuries and no feathers were scattered about. A search of the area failed to turn up the missing head or any other signs of death.

How did the hen die? How did she lose her head? Where was the head? Chickens are known to be cannibals, especially when confined to close quarters or when blood appears on an injured member of the flock. Could the hen have cut the comb on top of her head and the flock have pecked at it until she died and the pecking continued until the head was completely gone? Maybe a varmint like a fox or raccoon had killed the hen during the night while it roosted on the pole. But why did it only eat the head? Why didn't it drag the entire carcass into the woods or fields?

Whitney returned to the house to report her discovery to me and to enlist my help for carcass removal. She always took her responsibilities seriously, but handling dead animals was a job she liked to avoid. When she did remove a dead chicken, it was always held at arm's length with a scrunched-up expression on her face, even if there wasn't a decaying animal smell yet. She wanted to keep death as far away from her as possible. The task was always harder if it involved one of the chickens that she had given a name.

We returned to Chicken City for another investigation of the scene. There were no signs of holes in or under the fence. While I removed the dead hen, Whitney counted the flock. There were thirty-nine hens and roosters. As she gathered the day's egg production from the nests, she looked for signs of an intruder. There were none. No eggs seemed to be missing or broken. All that was missing was the head of one hen.

The next day, Whitney went to the chicken house apprehensively. She neither wanted the sadness of finding another dead hen or the surprise of walking in on the culprit. Fortunately, all was normal in the chicken house that afternoon and she went about the routine of putting out feed, checking their water trough, gathering eggs, and pausing for sweet talking with her friends. They didn't seem nervous or upset about yesterday's event, but

Whitney told them how concerned she was and encouraged them to be careful.

The next day the horror returned. Another headless hen was found. The scene was the same. Few clues seemed to point toward a likely culprit. It was hard to know what steps to take to prevent the killing, if I didn't know what was doing it. At church on Sunday, friends and neighbors with more years of experience in such matters were told the story and asked for suspects. The responses ranged from joining us in our puzzlement to being certain it was a possum. Also thrown into the line-up were fox, coyote, dogs, raccoons, and snakes. With all of those as possibilities, Whitney wasn't sure she wanted to go to the chicken house alone again.

There were no unusual signs for a few days, but one afternoon when she opened the door, there was a greater surprise: two headless chickens. And the day after, another one. By the end of the week, the flock was down to thirty. Egg production was declining rapidly and, at this rate, the entire flock would be gone if a solution wasn't found.

Friends at church greeted us, but didn't ask how we were doing. They wanted a chicken report! Their questions were a mixture of compassion for Whitney, curiosity over the novelty of it all, and the adventure of a mystery to be solved. It was no longer just Whitney's chickens in danger. It was a community problem. Everyone wanted in on the solution…even those who knew nothing about chickens or their predators.

No new suspects were added to the list and the suggested tactics shifted from discovery to protection. Maybe the chickens should be put into a big coop until the danger passed. Since the killings were happening at night, why not close the little door the chickens used to move in and out of the chicken house? After dark, when the chickens were on their roost, the door could be closed, and then opened again the next morning. It would be an added chore for Whitney, but it would at least isolate the source of the predator and hopefully eliminate the problem.

I rigged up a rope and pulley system for the chicken house door so she could stand outside the chicken pen and close or open the door easily. The morning and evening chores were repeated for several days and with success. With the predator's entrance apparently blocked, maybe it had decided to look elsewhere for prey. It was time to test for a new pattern and the door was left open one night. It didn't take long for the test results to come in. The next morning, another headless hen was found. It was as if an announcement had been made to all predators: "Come back, the door is open again."

The continuing carnage was reported at church again on Sunday. Speculation continued, and the suspects were narrowed down to raccoon and possum. The church treasurer, who also had a flock of chickens, told about a possum getting into his chickens and volunteered the use of his live animal trap. If the culprit with a taste for chicken heads could be caught and removed several miles away, the problem would probably be eliminated.

That afternoon his trap was put in the chicken house at Chicken City. Since the suspects had a taste for meat, some canned dog food was put into the trap for bait. After sundown, when all the chickens had gone to roost, the trap was set. Hopefully by the next morning the mystery would be solved and Chicken City would be safe again.

Before breakfast the next morning, Whitney went to check the trap. She got a surprise when she opened the chicken house door. There wasn't a headless chicken on the floor, but there was something moving inside the trap. It was a hen pacing back and forth, looking for a way out. She had eaten the dog food but definitely wasn't the perpetrator of the headless chicken crime. She just woke up when the sun came up and helped herself to a breakfast of dog food.

This scene repeated itself for several mornings until finally there was a new occurrence. This morning revealed the familiar chicken in the trap, but in addition, there was another headless chicken on the ground outside the trap. The next morning, it was the same: another chicken in the trap and a dead chicken outside the trap. The trap wasn't working. We were now down to eighteen chickens and no new clues or ideas.

The chicken report baffled the church crowd. Everyone just knew the trap would be the answer. Why were we catching chickens but no predators? Wrong bait? Bad timing? No one had new ideas. It finally came down to "maybe you ought to get out of the chicken business." Other than locking and unlocking the chicken house every day, nothing had worked.

Since the trap wasn't being effective, it was removed. The chickens probably missed their dog food treat. With no deterrent, the killings continued and in a few days the chicken count was fifteen. Maybe getting out of the chicken business was the best idea. Without a change, the predator would put Whitney out of business whether she wanted to be or not. With egg production a fraction of what it had been a few weeks ago, Whitney had already told everyone, except her most faithful customers, that there wouldn't be any fresh country eggs for a while.

One evening Whitney didn't get home until after sundown, but dutifully made her way to the chicken house to gather eggs in the dark. It was always a little spooky reaching into the nests when she couldn't see what was in there. Seldom would her hand touch anything but the familiar shape and texture of an egg, but her mind could imagine everything from snakes to rats.

She wasn't but a few yards from the chicken house when strange noises were heard. It wasn't chickens jostling for position on the roost pole. There was squawking, wing flapping, and bodies banging against the walls. Somebody or something was being caught in the act. Whitney approached the chicken house cautiously. She wasn't sure she wanted to confront whatever was creating the mayhem inside. She wasn't armed with anything but her egg basket and with all the noise inside there could be something in there of danger to her. Fortunately, she would be able to peek through a wire-covered opening and see what was going on without having to go into the chicken house. As she eased up to the opening, she didn't have to look far into the chicken house. Before she was ready to discover anything, she was face to face with a pair of eyes staring at her. She was only inches away from those big eyes blazing yellow like they were on fire. Only the wire separated her from them. Before she had time to analyze what she was seeing, she instinctively jerked her head away from the opening. All the blood drained from her already fair complexion as her heart skipped a few beats and she slid to the ground beside the chicken house. What had she just seen? The eyes were as big as hers. Maybe bigger. It definitely wasn't a chicken. Since those big eyes were on the same level as hers, Whitney thought a human being might be inside the chicken house. Before she could move from her crouched position below the window, there was more squawking and wing flapping. At the risk of staring face to face at those big eyes again, she slowly eased up to the window for another look. When her head reached the window, the big eyes were gone and she focused hers on the darkness inside. A quick scan only revealed the silhouettes of chickens trading positions on the roost poles. But when she looked down to the floor under the chickens, she saw those big blazing eyes looking at her again.

She didn't wait to analyze the situation further. She ran to the house as fast as her feet could carry her. In the ten second sprint to the house her mind moved as fast as her feet. What did those big eyes belong to? It was too small to be a human. But how did it get up to the window? Snake? No, too small. Dog, fox, coyote? No, too big. Possum or raccoon? Maybe, but the eyes were too big and they would have had to be standing on their hind legs. How could they get up to the window? There was a quick glance over her shoulder as she flew by the hackberry tree half way to the house. She wanted to be sure those big eyes weren't following her. They weren't, but she ran a little faster just in case.

She tore open the door and careened through the house to the living room, where I was sitting and reading. I already knew something was up by the way she ran through the house. Out of breath, she punched out a few words at a time. "There's something—in the—chicken house!—It's got—big eyes—and is walking—under the roost—poles!" All she said or did signaled an urgency to act quickly.

But in my usual calm, unflappable way, I began to ask questions like a detective investigating a homicide. Whitney didn't want to be interrogated and interrupted. "Come with me and let's see what it is. Maybe you can catch it." At this point, I wasn't so sure I wanted to catch a big-eyed chicken killer. What happens after you catch it? "Come on! Whitney said as she turned and ran out of the house. After months of mystery and failed attempts, she knew this was our best chance to solve it all.

The chicken house had become quiet as we approached it. Whatever Whitney had seen was either already gone or had wiped out the flock. We both peeked through the wire-covered window but didn't see anything but chickens cocking their heads from side to side. They were still on the lookout for danger. Not seeing the big eyes Whitney had reported, I went around the corner of the chicken house and pulled open the door enough to look inside. I still didn't see anything, so I reached inside and flipped the light switch. The sudden light created movement among the chickens on the roost poles and on the floor beneath. The two big eyes looked up at the two humans and everyone froze.

The predator wasn't human, nor was it any of the suspects that had been accused. It was an owl! More adept at flying than walking, it waddled about under the roost poles moving to a corner in retreat. Knowing this was our chance to catch the predator and put an end to the chicken massacre, I closed the door and quickly released the rope to close the small door the owl had used for entry. The owl was trapped, but how would it be captured?

Both Whitney and I knew the chaos that can occur if we tried to catch a chicken on the roost pole. If we were unsuccessful on the first attempt, wings would flap, feathers would fly and bodies would bounce off the walls all around us. We could imagine what putting a long-winged owl into the mix would mean. It could latch onto us with its sharp talons or rip our flesh with its powerful beak. We were content for now to let it huddle in a corner on the floor.

I knew one human would be less likely than two to excite the owl into flying. I had Whitney step outside the door and hold it closed,but be at the ready to open it in case I needed to escape. Seeing the door open, the owl made a move but only turned around in its corner. Just like the humans, it wasn't sure what to do in this strange situation. Seeing the owl's fear and uncertainty, I decided I might be able to open the feed room door on the back wall of the chicken house and give the owl a chance to retreat. I opened the door, moved along the wall to the owl's corner, and watched it waddle toward the feed room door. Built for flying, not walking, the owl seemed harmless in its waddle across the floor. It went hopping through the open door and I happily closed the door behind it. The owl wasn't fully captured, but at least it was separated from its midnight snacks. Peace was restored for all in Chicken City.

Whitney calmly turned to her flock where they stood facing her in three tiered rows on the roost poles, like a choir waiting for a word from the choir master. I didn't know if they understood any of her words, but her gentle voice had a calming effect on all of us and one by one they settled into their spot on the roost poles. After she reassured them that they were safe, with faint sounds of cooing chicken talk in the background, Whitney and I closed the chicken house door and made our way through the darkness back to the house.

A call to the area wildlife resources officer for advice revealed there weren't many options. The owl was a protected species and couldn't be killed. Turning it loose wouldn't solve the chicken killing problem. She volunteered to come the next morning with a net to capture the owl. She would take the owl twenty miles away to a wildlife preserve, which wasn't anywhere near a chicken city. Murder mystery solved.

Skating with the Andrews Sisters

Easter brings new life to the farm. The early spring rains warm the soil and the grass turns from brown to green again. It's also a time for kids to get an Easter gift of live bunnies, chickens, or ducks. When our daughter, Whitney, was eight, we bought three baby ducks for her. For a day or two we kept them in a cardboard box in the mud room of the house. She nearly rubbed off all their fuzzy down playing with them so much. She put a bowl of drinking water in their box, but they preferred standing in the bowl and splashing the water all over themselves and the box. It became obvious the soaked box wasn't going to last long, so we moved them to a wire floored chicken coop in the barn. With a heat lamp to keep them warm, more space to run around, and a bigger bowl to splash in, they were a happy trio.

In a couple of months their fuzzy down had been replaced with glistening feathers in an array of mottled colors. They had also gained names. Patty, Maxine, and Laverne. For a reason no one remembers, they were named after the Andrews Sisters. I'm sure it wasn't looks or voices. One warm Saturday afternoon in early June, Claudia, Whitney and I each picked up a duck from the coop and carried them down through the pasture to the pond. It was time to let them see what real water was like. Big water! We put them on the pond bank about a foot from the water's edge and watched, not sure what to expect. Apparently, they weren't sure what to expect either. There wasn't a mad dash to the water like there was to the bowl in the cardboard box or coop. This was no bowl they were seeing. One duck broke rank and ventured to the water's edge. Her bravery didn't last long, as she retreated to the security of the group before touching the water. They must have had a conference after the scouting report, because they all turned around in unison and marched into the water, stopping after a few inches to just stand in the water. The mud bottom must have felt different to their feet than standing in a bowl. They put their heads down, stuck their bills into the water and began a combination of drinking and playing with the water. They would rapidly open and close their bills, splashing water out of each side. One by one they walked deeper into the water until their feet no longer touched bottom and the buoyancy of their bodies kept them afloat. Continuing to move their feet as if walking, they got their first swimming lesson from Mother Nature. One duckling wasn't so sure about the situation and circled back to the bank. She shook the water off her oily feathers from head to tail and looked up at us as if to say, "Well, I'm through with that, what's next?"

With a sudden splash that startled all of us, one of the ducks revved up her feet into overdrive and headed toward the deeper water with such speed she almost went airborne. Not to be left behind and almost mimicking the actions of the first duckling, the others followed. They churned the water like three speed boats at the beginning of a race. We could sense their pleasure as they explored the expanse of the pond with their new freedom and new skills.

After a vigorous swim they circled back in our direction and stopped ten feet from the bank. One stuck her head under the water to look around. They all repeated the head bobbing routine a few times. In a one-upmanship fashion one duckling stuck her head into the water and kept going until her tail feathers pointed up to the sky. With her exhibition complete, she turned upright, shaking the water from her head and saying, "Try that girls!"

We watched their explorations and antics with the pleasure of knowing they were in their natural element and had been equipped with the instincts to make themselves at home. One by one they made their way back onto the bank and began their preening, feather oiling procedures.

Having brought some of their feed with us, we sprinkled it on the ground near them. They knew what to do with it too and began sucking it up like vacuum cleaners. They would begin eating what nature provided in and around the pond, but for a few days we would supplement their findings with the remaining commercial feed. After that, our feeding them would be limited to an occasional treat when we came to visit.

Throughout the summer and fall we made regular visits to the pond and watched them grow. We would usually take them something to eat. It might be some corn or chicken feed. We saved leftover bread for them. When we came for a visit, they would come from wherever they were on the pond to greet us. They might be sitting on the opposite bank, but would waddle to the water, slide in, and glide toward us. They hardly made a ripple in the water as they paddled along. When they reached our side of the pond, they wasted no time getting out of the pond and coming up the bank to where we stood. There was no stopping for preening. They waddled toward us and shook off the water as they came. Even though they were responding to our conditioning of them by bringing food, it always felt good to be greeted in a "we're glad to see you" fashion. It's one of the pleasures of having animals. They depend upon you for many of their basic needs and as they are driven to get those needs met by you, their dependency comes through as gratitude.

As we watched the trio eating our food offerings there was time to observe the changes that were taking place. We watched them continue to get bigger. We listened to the gradual maturing of their voices. And we noticed the changing colors of their feathers as they passed from adolescents into adults. One change came as a surprise. We had already been noticing the one we called Laverne was becoming more dominant in her relationships with Patty and Maxine. This seems to happen with most groups of animals, so it was no surprise. What we hadn't noticed was the very gradual curling of the feathers at the end of Laverne's tail. Her dominance wasn't just a matter of pecking order; she was a drake! With that curl of tail feathers and all the other male traits that came along with them, Laverne needed a name change. They weren't going to be the Andrews Sisters anymore either. Rather than break up the now familiar ring that Patty, Maxine, and Laverne had for us, we decided to let Laverne become Vern. So, from then on, it was Patty, Maxine, and Vern.

Winter seemed to come early that year. In December we were already having temperatures in the 20s. On Christmas Day we not only had snow for a white Christmas, but it had been cold enough, long enough for the pond to freeze over. As long as it stayed this cold, it meant a daily trip to the pond with an axe to cut a hole in the ice for the cattle to drink. It wasn't a pleasant chore, but essential for the cows to survive. The chore was offset with a winter pleasure. Kids from the neighborhood could skate on ponds. When I say skate, I don't actually mean skating like on a city ice rink. To start with, few had skates to wear. Everyone would go out onto the ice and slide around wearing their boots or shoes. Not having skates didn't seem to lessen the fun or the falls. The other difference was the size of the rink. Most ponds are far bigger than the confines of commercial rinks. You could skate in a straight line for a long distance and not have to keep going around in circles. As a boy, I remember going to Allen Shouse's house, especially when the snow and ice had cancelled schools the next day. Allen had about a twenty-five-acre lake that became our ice rink. A fire would be built on the shore where we could rest from our late-night skating games and warm ourselves as we planned the next game. Part of the fun was making up games as we went along. Rules or scores seldom mattered. We were just having fun and enjoying being together.

It was very unusual for it to be this cold on Christmas Day. After the traditional morning activities of opening presents and having a special breakfast, Whitney and I bundled up in several layers of shirts, sweatshirts, sweaters, and jackets. Moving wasn't easy, but we would stay warm. We were heading to the pond, where I would cut through the ice for the cows to get water. I wasn't sure, but thought it would be thick enough for us to skate. We also took some leftover biscuits for Patty, Maxine, and Vern.

There were about four inches of snow on the ground, so walking wasn't impossible, but a little more difficult than normal. The day before, when I checked on the ice, the upper end of the pond wasn't frozen for the first twenty feet. The pond was fed by a spring and its warm water flowing into the pond kept the upper end thawed most of the time. As low as the temperatures had dropped, I expected the pond to be frozen everywhere. The cows would need a hole cut for drinking.

Two things we saw when we got to the pond I'll never forget. Both were vignettes of nature's own struggling and coping with the cold. I expected to find the trio of ducks snuggled in the high grass on the far side of the pond, but they were in the water on the upper end of the pond. Ice had reduced the water's surface to a small two-foot hole where the ducks swam in tight circles. I don't know what kept that spot from freezing unless it was the activity of the ducks in it. The large expanse of the pond had closed in on the ducks a foot at time, until now there was hardly room for them to turn around. Their pool wasn't very big, but they were having a

Christmas swim. Vapor rose like steam from the water that was much warmer than the cold, dry air.

When they saw us, the circling stopped and one by one they came out of the little puddle of water and onto the ice. As usual, Vern was in the lead and about ten feet ahead of Patty and Maxine. He wanted to be the first to greet us and sample our Christmas gifts. The other two seemed to think that even though the water was cold, it was warmer than walking across the ice. Vern was half way to us with the others waddling behind, when he stopped and raised one foot. He stuck it out in front of him with its bottom facing our direction. After a couple of seconds, he put it back down, waddled a few steps, and raised the other foot in our direction. It was as if he were holding up the bottom of his orange feet for our inspection. In reality, they were getting cold from walking on the ice and holding them up let them warm up one at a time. He put the second foot back down after a few seconds and waddled in our direction again. The second memorable sight came when he stopped again. We expected to see a repeat raising of alternating feet. But this time he sat back on his tail feathers and raised both feet. He just sat there with both feet sticking out. He wasn't going anywhere until his feet warmed up! With Patty and Maxine duplicating the routine, we felt like we were watching a group of circus ducks show us their tricks. They eventually made it to shore and gobbled up their Christmas treats. I've never seen ducks sit with both feet in the air before or since.

I used my ax to cut a two-foot by three-foot hole in the ice for the cows to drink from. The ice was six or eight inches thick over most of the pond. As long as we stayed away from the upper end where the duck's puddle was, we could have a safe skate. I walked around the edges of the ice listening for the cracking or popping of thin ice. There wasn't a sound from the ice even as I moved toward the center on the pond. I've never come close to breaking through thin ice or been around when anyone did. Although years before, one of our 700-pound steers walked out onto the ice, fell through, and drowned. We didn't see it happen, but found its body when the ice thawed and assumed that's what had happened.

Whitney didn't have much skating expertise at this time, so I took her by the hand and we circled the pond together. As she gained confidence, she got her feet going faster than her body and started to fall. I kept her from going down by holding up on her arm. After a few circles around the pond, she began to get tired and complained of cold feet. I suggested she do like Vern. "Sit down and hold your feet up." She thought that sounded like a good idea. If it worked for Vern, it could work for her, so she lay down on her back and raised her feet up into the air. Like Vern, it did get her feet off the cold ice, but it wasn't very restful holding her feet up.

While she had her feet sticking up, I grabbed hold of them and told her to lock her knees and keep her legs stiff. I began pushing her around on the ice with her lying on her back. She especially liked it when I would come to the end of the pond and swing her around as I held onto her feet. Sometimes I would swing her around real fast, turn loose of her feet, and let her go sliding across the ice like a hockey puck. It reminded us of a game she and I played on the kitchen floor, when I would swing her around sliding on her back. We called it slip and slide. Since there was so much more room on the pond ice, this was more fun.

The ducks finished their Christmas treats while we had our fun on the ice. They then made their way back to their lake puddle and resumed their swimming. They looked so comfortable, we were tempted to stick our feet into their water, but knew we didn't have the feet for it. Just like they knew they didn't have the feet for ice skating.

Hypnotized Rooster

Twenty-one days after the hen's nesting began, a new chick poked his fuzzy little red head out from under the mother hen. The struggle to get out of the egg shell was just his first of many adventures. He had many enemies and much to learn.

His mother was the center of his world. She emanated warmth, protection, food, training, and direction. He had grown under her warmth for three weeks. Now with only his head exposed to the cold world outside, he appreciated her warmth for the first time. He couldn't imagine life outside the nest, but in a couple of hours he would find out. After days on the nest, the hen wouldn't be staying any longer than necessary. When she left, the chicks would follow, not to return to their hatching site. They would not stray far from the protection of the watchful eye and outstretched wings. Her scratching in the dirt and grass would uncover food. By imitating her, they would learn to scratch and discover for themselves. Her clucking would call back the strays and send a follow me signal when it was time to move along. When danger threatened, a higher-pitched, rapid clucking would bring all her chicks scurrying beneath the safety of those broad wings and warm body.

The chicks from this clutch came in an assortment of colors. Black, yellow, spotted, and two-toned. But in a few days, this red chick would have an experience that would make him even more different than the others.

On our farm lived a dog named Beau. We called him a strooch. Half stray, half pooch. He really was a mixed breed and resembled a mixture of German shepherd and Collie. Beau had been left on the farm by a newlywed couple who lived in the cottage and moved into an apartment that didn't allow dogs. Sometime in his early life, he must have been mistreated. He was shy and would cower at even a calm command. He was sweet with the kids and would let them do anything with him.

One hot summer afternoon, the hen and chicks had ranged a little further away from the chicken house and into the barn lot that ran alongside the chicken yard. The high weeds offered protection from both the heat of the day and the eyes of hawks flying overhead. They went about their routines of scratching, pecking, cheeping, and clucking. Beau came upon them suddenly, and the calm was turned into pandemonium. Beau went for the hen first. In the surprise, she reacted with a lurch to save her own life. The chicks scattered in all directions. The hen continued her evasion but squawked messages of anger at Beau, interspersed with clucks of comfort and direction to her scattered brood. With one eye on the dog, she searched the weeds for a place to reunite her chicks. Several had already found each other and were huddled in the corner of the barn lot fence. If the dog had seen them in the corner, they would have been doomed because he was between them and the hen. But he had seen two or three chicks running along the fence line away from the corner and trotted after them.

I was working nearby and my peace and quiet had been broken by the squawking. It was so loud and ongoing; I knew it was more than a momentary dispute among hens. I dropped my hoe and went to investigate. When I was within eyesight of the barn lot, I saw Beau trotting along the fence with a mad hen in pursuit. This scene seemed to be in reverse. Shouldn't the dog be chasing the hen? Getting closer, I saw it was a hen chasing a dog chasing chicks. This made more sense. In a flash, the dog reached down and gobbled up a chick. I yelled, "Beau!" The dog stopped in his tracks and looked at me. As I ran toward Beau, the dog knew he was in trouble and gave a pitiful "I'm caught" look. He changed from the aggressor to the accused. The dog was motionless, but the hen and chicks scrambled to reunite. When I got close to Beau, the dog looked up at me sheepishly. Sticking out of each side of Beau's mouth was the leg of a chick. I yelled at Beau again. Beau opened his mouth and out tumbled the red chick. It dropped lifelessly to the ground. I knew it had been crushed in the mouth of the dog.

Beau and I were looking down at what had almost been an appetizer for the dog, when its legs began kicking spasmodically. It wasn't dead, but would need to be put out of its misery. Before I could climb the fence and do the job, the red chick turned over, raised its head, and tried to stand. It wasn't successful and collapsed. It didn't stay down long though. The second attempt to stand was more coordinated, but he still didn't look normal. He seemed unable to control his head as it flopped from side to side, as if his neck was broken. After a few seconds, his head began to stabilize, and he tried running away from Beau, who was watching curiously. The chick's world must have been spinning because he wobbled fore and aft as if on the deck of a ship in a storm. First his head would be out in front and his feet running so quickly one would think he was about to run away. Then as he leaned too far forward and was about to fall on his face, he pulled his head back, so far back it stopped his forward momentum and he began moving backwards. Forward and back, forward and back, he moved in dizziness. He was alive. He was moving. But the world he had returned to wouldn't stand still. After a few more lurches, he seemed to find his sea legs and scurried to his mother's side.

Neither Beau nor I could believe what we had seen, but that was life on the farm. Birth, life, death, and resurrection all on a hot summer afternoon. Telling the family about the episode, I decided the red chick should be named Jonah since he had faced a certain death only to be brought up out of the mouth of the dog to live on.

Jonah grew up with all the other chicks in a normal fashion, but anytime we saw them, we always located Jonah and remembered his experience with Beau.

Whenever it came time to thin the flock for the dumpling pot, there was never any doubt that Jonah had earned the right to live out his life to its fullest. One day he might become the top rooster in the pecking order. If it was dependent upon grit, fight and determination, there would be no doubt.

By mid-December, the chicks were nearly full-grown. The young roosters squared off for short fights that would settle the pecking order and determine which rooster would dominate the flock. For now, the fights only settled the bottom end of the pecking order because the big old rooster still ruled the roost.

The young pullets began laying their first eggs, which were about half the size of a regular hen egg. It would take a lot of them to serve breakfast for us. Imagine the surprise of a neighbor who gets a gift of a dozen eggs and opens the carton to find twelve tiny pullet eggs.

As we made our way in and out of the chicken house to check on the flock, gather eggs, or fill the feeder, we began noticing a gradual change in Jonah. Rather than scattering with the flock in fear, Jonah would approach us with a confident curiosity. At first, we thought it was the beginning of a friendly relationship, but in a few days, it was apparent that the advances were aggressive, not friendly. First it was just a darting back and forth in a series of advances and retreats. Then the retreats became fewer as he began to stand his ground. When the feathers on his neck began to stick out, it was a sure sign he had an attack in mind. The neck feathers sticking out were to give his opponent an impression that he was bigger than he really was and to intimidate him enough that an all-out fight would not be necessary.

At first, a push of my leg was enough to end Jonah's aggressive encounter, but one day Jonah responded to the push with a shove of his own. He did not retreat. He ran at me and made a flying feet-first leap at my leg. It took me by surprise. Jonah had decided to be a fighting cock.

One by one, we discovered Jonah's new attitude. Egg gathering became an exercise of find Jonah, find the eggs, watch Jonah, gather the eggs, check Jonah, finish the egg gathering, and slip out without coming to a face-off with Jonah. Unless someone had bare legs, Jonah wasn't going to hurt them with his flogging. His spurs hadn't developed and his toenails were his only weapon. But everyone could still imagine him jumping into their face and scratching their eyes out. That fear was all Jonah needed to dominate even the humans.

By spring, Jonah was full-grown and had even conquered the big old rooster. He was first in the pecking order and strutted around the chicken pen with a pride and confidence that let everyone know this was his flock now.

Visiting family and friends had to be warned about Jonah's aggression. Most were content to just watch the flock from outside the fence, while one of the family slipped in and out to gather the eggs. Jonah didn't let the fence stop him. He squared off with visitors as if the fence wasn't there. If ruffling the neck feathers didn't back them up, he would leap at the fence and that made the visitors scatter.

In early summer, while we were away, Rich, who lived in our cottage, gathered the eggs. He had already heard stories of Jonah's aggressive attitude and experienced some of it first-hand. His father was visiting from Cincinnati and enjoyed doing farm chores. Knowing what would happen, Rich sent him to the chicken house to gather the eggs. Nothing was said about Jonah's attitude. Rich secretly watched from a distance, as his dad went through the chicken pen gate and locked it behind him. He disappeared into the chicken house. As his bad luck would have it, Jonah was in the chicken house. A confrontation was inevitable, but the man never saw it coming. He just waded through chickens on his way to the nests and went about his business of gathering eggs. He hadn't even gotten the first egg when Jonah hit him from behind. Turning around to see what had hit him, he was standing face-to-face with a very mad, ruffled-neck, red feathered rooster. Before he could evaluate the situation, Jonah hit him again. The startled man forgot about the eggs he had come after and took a step back. Jonah knew he was winning and hit him again. All Rich could hear from a distance was the flapping of wings, cackling of hens, and his father's wordless expressions of surprise.

"Acht, acht, acht!" This city slicker couldn't find words or actions to fend off Jonah's attacks as he backed out of the chicken house. Jonah hounded him all the way to the gate. Rich laughed and watched his dad fumble with the gate latch as Jonah unmercifully stalked him from behind. Finally, able to unlatch the gate, he opened it just enough to squeeze through as Jonah made one last charge. His brief egg gathering having ended, he asked his son, "Why didn't you tell me about that dammed rooster!"

When we returned from our vacation, everyone had a big laugh as we heard about the father's episode with Jonah. As the weeks went by, Jonah's cockiness only increased. Some of us refused to take part in egg gathering chores. As Jonah's dominance spread, it became obvious that a remedy was needed. While the remedy came unexpectedly, it was recognized immediately.

Jerry was the previous cottage tenant and returned to the farm for a visit. As he told about life on the new farm he was renting, one of the things he said he missed was having chickens around. He didn't have a chicken pen or house, but would be content to let them roost in his barn and roam free. The hens would make nests for themselves. I offered to give Jerry some of our hens and a rooster to start his flock. Naturally, Jonah was the candidate for Jerry's new rooster. Not wanting to get Jerry into something unexpected, he was told all the stories of Jonah's life history. Jerry just chuckled and wasn't fazed by Jonah's reported aggression. Everyone decided there was no better time than the present moment to catch the chickens and send them with Jerry to their new home.

As we walked to the chicken pen, we discussed how to catch, secure, and transport the chickens. Since a coop wasn't available, Jerry suggested tying their legs together with hay twine and putting them in the bed of his truck. As we easily caught a few hens, tied their feet, and laid them in a row on the ground, I told a story about a frontier family. They had moved so many times, that when the chickens saw them putting furniture into the wagon, the chickens would lay down beside the wagon and cross their legs waiting to be tied and loaded.

Next came the dilemma. How would we catch Jonah? He wouldn't stand still long enough to be snared with the chicken catcher, which was a stick about four feet long with a hook shaped wire on one end to slip around the leg of an unsuspecting chicken. Cornering Jonah in the chicken house would only bring on a fight with plenty of squawking and feathers flying. Jerry confidently volunteered to catch Jonah which relieved everyone else. How would Jerry catch Jonah? "I'll hypnotize him," Jerry said. No one had ever heard of anything like that before and generally disbelieved it could be done. Surely Jerry planned to do something similar to hypnotism. What would he do, swing a pocket watch in front of Jonah and slowly say, "You're getting sleepy, sleepy?" Before anyone could ask or Jerry offered to explain, he was on his way across the pen toward the remaining flock. Jonah didn't have to be singled out. He separated himself from the flock and met Jerry's advance. When Jerry got about six feet from Jonah, the rooster's red neck feathers were already ruffled. It was the familiar scene of the beginning of a fight. Everyone knew a flying, feet first leap from Jonah would be next. But before Jonah could leap, Jerry crouched down in front of him and they were eye to eye. Jonah must have thought this was going to be easier than imagined. He'd never had such close access to someone's face before. Jonah scratched with both feet as if to get traction for his leap. Jerry lifted both hands and extended them with his palms facing the rooster. It was a move that obviously surprised Jonah, and he stood up from his pre-leaping crouch. Jerry began moving his hands from side to side. His left and right hands crossed as they moved rhythmically in opposite directions. Jonah's head began watching Jerry's hands, first the left, then the right and back again. With

Jonah watching the hands, Jerry moved closer to Jonah. At first Jonah thought he would attack the advancing opponent, but he wasn't sure where to attack him. As Jerry inched closer with his hands still crisscrossing slowly, Jonah backed up to give himself more time to analyze the situation. Jonah was perplexed. He moved first forward, looking for a point of attack, and then back in puzzlement. Jerry kept moving forward until they could almost touch each other. Jonah finally broke and turned to retreat. When he did, Jerry swiftly reached out and caught Jonah's legs from behind. It was all over in an instant. The defeated and captured Jonah hung upside down by Jerry's side on his way to have his feet tied like the hens and take his place in the row of other prisoners.

Everyone was impressed with Jerry's skill and heaped him with praise. Then we all wanted to know how he had done it, even though we had all just witnessed it. He explained that the rooster wasn't hypnotized, but confused. The rooster was watching the hands and looking for a place to attack. The constant crisscrossing of the hands gave him a target that moved before he could take aim and leap.

For weeks after Jerry loaded the chickens into his truck and hauled them to his new farm, everyone repeated the story of how Jerry hypnotized the rooster. No word came from Jerry on how his new flock had adjusted to their new home until one Sunday afternoon at the end of the summer when he stopped for a visit. He told of how the chickens spent the night roosting on a pole in the barn and how the hens had made a nest on a pile of old feed sacks in the stable. Of course, what everyone wanted to hear was another story from the legendary Jonah. Jerry only had one story to tell.

He said he was standing on his back porch late one afternoon and could see Jonah and several hens between the house and barn. They were calmly going about their routine of scratching and pecking the ground in search for food. Jerry looked up to see a hawk circling overhead. It swooped down among the chickens and landed on the back of the red rooster. In an instant the hawk sunk his claws into the back of the unsuspecting Jonah and with a flap of his wings lifted Jonah off the ground. Slowly and laboriously, but with certainty, the hawk gained altitude as it carried its prey in its claws. Jerry watched as they went out of sight, never to be seen again.

Sneaker

When the mailman delivers a box with twenty-five day-old chicks, no one needs to tell anyone what's inside the box. The cheep cheep cheep sounds from inside the box is their announcement that they are ready for their first drink of water and first meal. For a few weeks they will all look and sound alike. They are a ball of fuzz from head to tail. But when their feathers begin to appear, one notices about half of them have tail feathers that point up, while the other's tail feathers continue straight back. In a few more weeks, those with the tail feathers pointing up will start making a strange sound. It's not a cheep, cluck, or crow. It sounds more like a soprano singer with laryngitis. It's a young rooster's first attempt to crow. He sounds anything but proud and king of the roost. He throws back his head as he takes a deep breath. When he moves his head forward and exhales with a wide-open mouth, it is supposed to come out as a crow, but it comes out as a screech, then a gurgle, and ends with a whimper. It will take some growing up and practice before he can perform a genuine cock-a-doodle-do or Er-a-er-a-errrrr.

Most folks would be happy to have a box of chicks with one or two roosters and the rest hens to start their flock. Hatcheries sell chicks three ways. Commercial farms raising broilers buy all roosters for their meat. Laying operations buy all hens for their eggs. The rest of us buy a straight run hatch, which is how God makes them, about half roosters and half hens.

Since we don't need all of those roosters, we have to decide what to do with them when they mature. Some end up in the frying pan which involves scaring the kids and shocking the city folks as beheaded chickens flop around the yard after one wrings their necks. Then there is the scalding and plucking the feathers. Few like the smell of wet chicken feathers and it is quite tedious to get every last tiny feather. The evisceration is a smelly thing that reveals more about a chicken than most want to know. Even with all of this, there aren't any chicken fingers or nuggets in sight. With Tyson doing all the work for us, few are left who have ever cut up a whole chicken.

The only other alternative for disposing of your extra roosters is to find someone who wants a rooster and give them away. It may work with puppies, but I doubt one would give away many roosters standing in front of Wal-Mart. Regardless of my strategy, I've only been successful twice in giving away my roosters. The first time was to a neighbor who had a flock of chickens and a hawk had flown away with his rooster. That must have been a sight to see that hawk swoop down out of the sky, grab a bird three times his size, and fly away. The second rooster gift was at a retirement roast for a colleague at work. What better thing to take to a roast than a rooster! Actually, we roasted my colleague, not the rooster. He grew up on a farm, but had lived in the city after finishing school and had missed waking up at dawn to the sound of a crowing rooster. When I heard him say that, I knew I could help him and myself with the gift of a rooster. I would have thought he would have preferred to retire to quiet mornings of sleeping later. I suspect the neighbors in his subdivision would too. He never commented on the rooster, so I don't know how that worked out.

One summer all of our roosters had met their destiny except two, and since they were competing quite aggressively for domination of the chicken yard, the hens, and especially the other roosters. I decided to take them down the hill to our country inn. I figured our guests from the city would be fascinated to see and hear a real live rooster.

I was right. They were a big hit with everyone except Bettie, the housekeeper. Since I didn't want the roosters to think of the inn as the center of their residence, late one afternoon when they had gone to their roost poles in the chicken house, I caught them and took them to the barn, which was about a hundred yards from the inn. I set each one on a brace pole in the barn, and they settled down as if they had been there all their lives. I figured they would hang around the barn most of the time and guests could see and hear them when guests were on a walk. That plan didn't work out. The next day the roosters had found their way to the inn and the next night they were roosting on the arms of a bench on the porch of a guest cottage. After a few days of the roosters leaving their droppings on the porch, Bettie said, "One of us has to go, either it's me or those roosters." I didn't want Bettie to go, so I moved the roosters again. This time to another barn nearer the inn. It was only their dropping on the porch that had caused a problem and I thought this would solve things. It didn't. The next night the roosters were back on the porch bench to roost. That day Bettie removed the bench, but the roosters found a bench on another porch. It was Bettie who found the neighbor who had lost his rooster to a hawk and she cut her rooster problem in half. None of the guests had complaints about the rooster. Just the reverse, they liked seeing him. Bettie must have made cleaning the droppings from their porch her first morning chore.

The rooster soon learned the easy sources of food. Tricia, our niece from New Orleans, liked to sit on the sidewalk after breakfast and let the rooster eat bits of leftover biscuit from her hand. A businessman from New York could hardly eat his breakfast for watching the rooster sitting on the dining room window ledge near his table. It was like he was at the zoo and looking at an exotic animal. He commented on the rooster's every feature and move. "Look how he cocks his head to look at me," he said.

Kids seemed to enjoy our chickens the most. When our friends, the Willis family, came for a weekend, their five and seven-year-olds, Amy Beth and Andrew, got to collect eggs one afternoon and eat them for breakfast the next morning. Now that's farm fresh eggs! The rooster showed up on the window ledge at breakfast hoping for a leftover biscuit. Andrew asked me, "What's that rooster's name?" I told him the rooster was new around here and we hadn't named him yet. "Can I name him," he asked. "Of course, you can," I replied. Off Andrew went to get acquainted with the rooster. For the rest of the day, every time I saw Andrew, he was following the rooster. I'm sure the rooster would be glad when this boy returned home.

Before breakfast the next morning Andrew walked up to me, stood as tall as he could, and proudly announced, "I've named that rooster. I named him Sneaker." I said, "That's a funny name for a rooster. Why did you name him Sneaker?" Andrew said, "I've been watching him." (That was an understatement!) "When he walks, he walks like this." Andrew lowered his head and leaned forward slightly, putting one hand in front of the other gently to imitate the rooster walking. "He walks like he is sneaking up on somebody," Andrew said. Of course, I thought, all birds walk like that. They are very tentative when they put their feet down. It took a small boy following a rooster all day to point out the obvious. Sneaker was a perfect name.

The next weekend Bettie got some unexpected help with her rooster eradication project. Claudia and I were working outside around the inn, and we heard a couple of squawks, then silence. When we went to investigate, we discovered our six-month-old Golden Retriever, Lane, standing beside a dead rooster. It was Lane's first trip to the inn. She had walked the half mile from our house to find us. Obviously, she found Sneaker along the way and did instinctively what all bird dogs do. The roosters were now all gone, but at least we got to keep Bettie.

Thanksgiving Survivors

Turkeys, like several other bird species, bond with the first big creature they see. This works really well when the first thing they see is their mother turkey hen. It becomes strange if they see a duck, pig, or dog. When I bought our first turkeys from a mail order catalog, I wondered what these turkeys would see first to create their lifetime bond. Since they were hatched in a mechanical incubator, maybe they would be like robots and follow machines around. Or would they see the postmaster through the ventilation holes in their cardboard shopping box and bond with him? I hoped not because they would have a busy life following him all over the neighborhood on his rounds six days a week.

Our turkey chicks must have saved their bonding for the moment we lifted the lid off their shipping box. They looked like fuzzy giraffes with their long necks and legs as they turned their heads to one side and gave us a one-eyed first inspection. I don't know anything about bird vision, but have observed them giving things a stare with one eye, when they are trying to figure out a strange new sight. We probably looked as strange to them as they did to us.

Our chicks usually arrive the week before Easter so we can have them on display for the kids at our restaurant. The baby chickens and ducks are the most popular. The kids handle them so much that I'm surprised the chick fuzz isn't worn off by the end of the day. The prehistoric looking turkey chicks aren't as cute and don't get handled as much, which probably suits them just fine.

We have learned from previous batches of poultry that it is best to wait a few weeks before we give names to them. One year we named our three ducks, Patty, Maxine, and Laverne after the Andrews Sisters, only later to have to change Laverne's name to Vern when she turned out to be a male. Our turkeys turned out to be two males and a female, and we gave them the good old Southern names, Bobbie, Billy, and Suzie.

Our Great Dane had grown up with poultry running around, so when the turkey chicks were big enough to live without their heat lamp, we let them have the run of the lawn. This didn't work as well when our next dog was a Golden Retriever and reminded us that she was bred to be a bird dog.

From our breakfast room window, we had a view of the turkeys foraging on the lawn. They roamed about leisurely scratching then pecking at the bugs and seeds they uncovered. Only the sight of a live grasshopper trying to escape would change their pace as they rushed to catch it.

As soon as one of us stepped outside the kitchen door, the turkeys would stop their foraging and head our way. They weren't expecting a food treat like a dog. They just wanted to be close to their parents. They would then follow us wherever we went. If I went to the garage to get some tools, they followed to make sure I got the right ones. If our daughter went to the garden for fresh tomatoes, they followed to make sure she got the best ones. If my wife walked to the car to drive to town, they escorted her to the car like a team of security guards and then gave her a "why can't we go" look as she drove away.

The biggest delight they gave us was when we returned home. As our car pulled to a stop in front of the house, we could see the three of them trotting in our direction to welcome us home. Everyone likes it when they arrive home and someone is glad to see us. For some it is expressed with a wag of the dog's tail, but for us it was the turkeys trotting toward us. They couldn't wait to welcome us home.

When children get in a hurry, they run. Ducks waddle. Turkeys trot. Their trot is a blend of a chicken running and a duck waddling. A grown gobbler may weigh thirty pounds and it takes some real maneuvering to keep his weight balanced over two legs and headed in the chosen direction.

The turkey trot expression has been popular for over a hundred years, but none of its associations seem to resemble our turkey's moves. In the early 1900s people were dancing their four hopping sideways steps to ragtime music. The turkey trot dance was pushed out by the foxtrot which didn't look like a fox's moves either. In the sixties Little Eva tried to bring back the turkey trot, but it was no match for the mashed potato.

I've never attended a turkey trot race, but every community seems to have one around the Thanksgiving holidays. I could picture some of the runners looking like our turkeys as they trotted along. If our turkeys entered the race and won, they would be happy to pose for a picture with the beauty queen presenting the trophy. One of our favorite pictures is of our niece, Ashley, posed with the three turkeys. The boys, Bobby and Billy, were lined up on the front row. The girls, Suzie and Ashley, were on the back row.

This occasion wasn't Ashley's first with turkeys. The first Thanksgiving after we married Ashley and her family joined us on the farm for a true country Thanksgiving. There weren't any pilgrims or Indians at this Thanksgiving, but the family got to witness all that was required to get the Thanksgiving turkey from field to table. Ashley and her brother, William, watched from a distance as the hatchet did its job. They didn't stick around for the feather plucking. Turkey slaughtering wasn't one of William's Boy Scout badges. The whole family gathered around the kitchen sink as the last of its pin feathers was removed. I'm sure the Thanksgiving turkey on the table was never the same for them. They would have been satisfied to just have it show up on the table with a brown roasted glow.

Most of our turkeys never made it to our dinner table. They succumbed to old age or invasions of neighborhood dogs. Each Thanksgiving Bobbie, Billy, or Suzie would take their turn going back to our restaurant to be in a cage and on display. The sign on their cage read, "I'm a Survivor. Happy Thanksgiving!"

Picky Peacocks

How much does a pair of peacocks cost? My first pair I got in a trade for two rolls of hay. My second pair I received as payment for officiating a funeral. My previous experiences with peacocks had always been from a distance and short term. It hadn't even occurred to me that a peacock was actually the male of pea fowl and that every pea cock had a pea hen. I had usually seen them on display in large pens and thought they added a touch of elegance to their surroundings. I guess I also thought all pea fowl had a beautiful array of tail plumage they could fan out for show. I then learned they, like the rest of the bird kingdom, reserved the beautiful plumage for the males of their species. The pea hen is a drab brown bird with no colorful tail feathers to show off. This was just the beginning of what I was about to learn on the subject of pea fowl cocks.

I had sold a bull to my neighbor, Joe Bowers, and when I made the delivery, it was my first time to visit his farm. He had been buying cattle and hay from me for several years, but he had always picked them up at my farm. Pulling into his barn lot, I noticed a peacock (I do mean cock, not fowl) strutting around the lot. As I got out of the truck, the peacock spread his tail feathers as if on cue. The sunlight turned his feathers into a Technicolor show of radiant blue, teal green and pure gold. Even the onyx black eyes on the ends of the tail feathers sparkled in the sun. If someone wants an example of blue, show them the neck feathers of a peacock. I thought he was just showing off his beautiful tail feathers for me. I later learned I was invading his territory and he was spreading his tail feathers in a defensive move. Instead of his being a little bird about one foot wide with one head, he was now five feet wide and appeared to have a head at the end of every feather. If an attacker was brave enough to take on such a big bird, it wouldn't know which head to strike. Most likely it would end up with only a mouth full of feathers as it chose to attack every head but the real one.

When Joe opened the barn door, I saw that this wasn't his only peafowl. There must have been fifteen or twenty peacocks and peahens walking in the barn alleyway, scurrying out of stalls, or jumping down from their perch on a stall wall. I asked Joe where he got all of these peacocks. With a response reminiscent of Noah's ark, he said they all started with just one pair. I told him I had been thinking about getting some peacocks and he said he could fix me up. I had no idea what a peacock was worth, but the fifty-dollar price he named for a pair seemed reasonable to me. Always looking to make a trade, since he had an abundance of peacocks and I had an abundance of hay, I proposed we trade a pair of peacocks for two rolls of hay. A deal was made and a few minutes later we were stalking peafowl.

Like catching most animals, feed was a good lure for getting close enough to catch them. I had never caught a peacock, but had often caught chickens by the leg with a long stick that had a sturdy wire attached. The V-shaped wire loop would slide over one of their legs, but when one pulled on it, the chicken's foot wouldn't slip through the loop and they were caught. When I mentioned my chicken catcher, Joe said he didn't have one and besides, we had to be careful when catching a peacock not to break their legs. He sounded like he was talking from experience when he told me, if a peacock was struggling to get away while we were holding them by their long thin legs, the torque of their big bodies could snap their legs. He tossed some cracked corn into a fully enclosed feed room and several peafowl went inside for the unexpected feast and an unexpected capture. I held the door closed while Joe was inside for the catch. I listened for the unwanted sound of legs snapping like a dry tree limb, but what I heard was even more alarming. I heard a high-pitched cry that sounded like a woman in great distress. I would hear that same cry many times in the months to come. It wasn't a cry from Joe or a woman he was hiding in the feed room. It was the first peafowl he snared. He pushed the door open just enough to slide a peahen through the opening and into my arms, before he went back to his stalking of a peacock. The peahen was actually quite calm as I gently but firmly held her legs with one hand and wrapped my other arm around her wings and body. In a moment there was another cry of a woman in distress, but I knew it was a male in distress. Joe emerged from the feed room holding a not-so-proud peacock that from head to tail was almost as long as Joe was tall, and with no disrespect to Joe, a whole lot more handsome.

We put the pair of peafowl into my cattle trailer, and I drove home happy with the trade of two rolls of hay for two peafowl. When I arrived home, I pulled my trailer into the barn lot and swung open the trailer door. Looking first left, then right several times, the peafowl hopped out of the trailer and trotted across the barn lot to the barn.

I knew nothing particular about the care and feeding of peafowl. My college class in poultry science didn't include peafowl. But Joe had said they were pretty self-sufficient. In spite of their self-sufficiency, Joe had warned me that peafowl were picky about where they lived and wouldn't stick around if they didn't like their surroundings. On the other hand, it was considered good luck if a peacock chose to come and live with you. I would come to appreciate the full significance of this characteristic some years later. Joe said they would forage around the barn and pasture for food and often chose a high tree limb for roosting, instead of inside the barn. My new pair first chose to roost on a high support beam in the barn, but later moved out to a maple tree in the barn lot. It was fascinating to watch them at dusk jump to get airborne and then fly their big 747 airliner bodies into the tree for their nightly roost. The next morning, shortly after sunrise, they would jump off their perch and slowly glide on a flight path that brought them gracefully to the ground twenty or thirty yards from the tree. Roosting at night in the tree kept them safe from predators and gave them a high perspective on what was going on around the property. I didn't know I had acquired a new pair of security guards until one evening I heard that same cry of a woman in distress that I had heard the day I got them. When a stranger, man or beast, came onto the property during the night, a peacock would give a warning cry as effective as any security system alarm.

During the day the peacocks moved gracefully around the barn lot, minding their own business. They weren't aggressive; instead they always seemed wary of people and other animals. When someone appeared, the cock would display his tail feathers, as if on cue. Actually, it was in response to his inborn defensive cue. Regardless of the trigger, it gave us the show of tail feathers that was the main benefit of having peacocks. Another benefit of the showy tail feathers was after they had fallen out of the cock's tail. From time to time, we would pick up tail feathers from the barn lot and put them on display in a vase in the house. Friends also felt special when we gave them a handful of feathers for their house.

Not all of the peacock's attributes or habits were enjoyable. After a few months we were surprised one morning to walk past our front door and see the peacock standing on the porch. He was looking through the full view glass storm door, as if he was inspecting the contents of the living room for a future burglary. He was cocking (now there's an interesting word to use) his head and moving it from side to side to get different angles of the room. He moved forward a little and then back again, but stayed on the porch and focused on the door. It was nice to be able to stand on the other side of the glass door and get such a close-up view of his fabulous feathers. He usually wouldn't let anyone get anywhere nearly this close. We could see the detail of his dark eyes and the layers of royal blue feathers on his long silky neck. Seeing us didn't alarm him and he would stay on the porch looking at the door for ten or fifteen minutes. On other days we could tell he had been back to the front door because he had left presents of Hershey Kisses poop piles. That's not a greeting we wanted for welcoming human guests at our front door.

We soon learned the peacock wasn't at our front door inspecting the contents of our living room or waiting for an opportunity or invitation to come inside. No doubt, one day he walked past our front door on his way to eat bugs from our lawn, when he caught some movement out of the corner of his eye. Stopping and looking closer, sure enough, there it was: the most threatening creature of all, another peacock! Of course, he was just seeing his reflection in the door glass. No one had explained mirrors or reflections to him in peacock school, so there was nothing to do but go onto the porch and approach this intruder, lest he claim all the peahens in the barnyard for his own.

Our front door and living room weren't the only places the cock found us keeping rival peacocks. He also found them in our car. He could see "them" in the car widows and would hop onto the car's hood for a closer look. Unfortunately, we started finding his Hershey Kisses on our cars and truck's hood and roof. But worse than the poop were the scratches from his razor-sharp toe nails that were also left behind. We became concerned that when our guests would park their cars in the driveway. Naturally our peacock would find they had brought a handsome peacock with them and felt required to hop onto their car to protect his turf. To our knowledge, none of guests ever got scratches on their cars or opened their car doors to drop off the peacocks they brought. We gained peafowl the natural way.

In the late spring I noticed our peahen was missing. The cock was still in the barn lot and looked normal. My first thought was a dog or fox had killed the hen. I looked around for leftover brown feathers, but found none. After about a week of keeping my eyes open for signs of the departed hen, I got my answer to her disappearance. I was on my tractor and bush hogging the weeds from the area where I fed hay to the cows during the winter. The weeds had grown waist high. Just in front of the tractor tire, the peahen flew away, startling me so much I stopped the tractor to catch my breath. It's a good thing I did because when I looked down at the spot where the hen had been sitting, I saw little peacock chicks scrambling in every direction. The hen had used the high weeds as a place to create a nest and hatch her brood. So, the weeks I had been missing the hen, she was laying her eggs and sitting on them the necessary twenty-eight days for incubation. If she hadn't flown away the moment she did, my tractor tire or the bush hog would have made a sad ending for her and the new chicks. I just turned off the tractor and let the situation calm itself down. It took twenty or thirty minutes of calling from the hen and cheeping from the chicks for the brood to all get reunited with the hen. I counted the chicks and came up with eight. That's $200 or eight rolls of hay I nearly ran over.

It only took another spring with that kind of production for me to realize why Joe had a barn lot full of peacocks. So, did I! The next spring the hen chose a similar spot in the feed lot for her nesting. This time, when she disappeared, I knew what to expect and went looking for likely nesting places with high weeds. She was just about twenty feet beyond the barn lot fence. Apparently, she liked the safety and privacy but still didn't want to go too far away. I had no intention of getting into the peacock business, but here I was now with a flock of eighteen. It was time for either a $50 a pair sale or some bartering for two rolls of hay.

The following winter I came across my first opportunity for flock reduction. We were visiting with our friends, Floyd and Anne Craig, at a birthday party they were hosting for another mutual friend. At the dinner table, Floyd told one of his true stories with the graphic detail his stories always have. We had just experienced one of our infamous Middle Tennessee ice storms, which starts off as rain and turns to sleet as the temperature drops. The result is everything outdoors is covered with as much as an inch of ice. Apart from the beauty of it creating a crystal palace, the result is mostly destructive, hazardous, and even life threatening.

Life threatening was the point of Floyd's story. We had all heard during the summer a declaration by the Craigs of their indisputable sign of good luck. A lone peacock had wandered onto their property and taken up residence in their stable. They knew their luck was especially strong, because the peacock chose them in spite of having to share the stable with the Craigs' dog. The two didn't get along, but their confrontations always ended with the peafowl hopping up onto a fence or wall and out of reach or flying away from the determined, but frustrated dog. The ice storm changed the equation and, without Floyd's graphic description of an iced covered peacock, one gets the picture. The Craigs' luck had changed. They still had a dog, but no peacock.

Unlucky? No peacock? I could make them lucky again! Or so I thought. Floyd's birthday was February 25 and I would give him the gift of a pair of peafowl. I knew he would appreciate them more than two rolls of hay, and I sure wasn't going to give him $50. At his party, I didn't actually present him with the peacocks, but when he opened his package, he was puzzled to find only one peacock feather. I quickly explained it represented two real peacocks that I would deliver soon. Floyd felt lucky again.

On the arranged delivery day, the Craigs were going to be out of town for the weekend, so we agreed I would put the peafowl in their horse trailer. I was a bit over-equipped, but just as I had done the day, I brought my first peafowl home from Joe Bowers farm, I "loaded" Floyd's peafowl into my cattle trailer and hauled them to the other end of the county. I backed my trailer up to the back of their trailer, slid open the doors, and the transfer was completed easily. I left the peafowl in their temporary home with plenty of water and cracked corn to last beyond the Craig's return day. They were also safe inside from the dog. I got a call on Sunday night that the Craigs had returned and found their peacocks, safe, comfortable, and contented. At least they looked contented as they walked around the spacious horse trailer in their elegant, but wary way.

The next evening, I received a call from the sad and apparently unlucky Floyd. After putting their dog into the house, the Craigs gathered around the horse trailer for the ceremonious opening of the door and release of the peacocks into their new stable lot home. And ceremonious it was, especially as Floyd told it. When Floyd opened the door, the peafowl pair walked calmly to the edge of the trailer and looked left and right to make sure the coast was clear before they hopped down to the ground. They hopped, but it wasn't down to the ground. They hopped up into the air and sailed fifty yards down the hill to the edge of the Craig's property. When they landed, they kept running, and to this day the Craigs haven't seen as much as a feather since. No doubt they went to a neighbor's farm to live and make someone else feel lucky. Should I make up for Floyd's loss by giving him $50 or at least two rolls of hay? Maybe another pair of peacocks to test his luck.

Like the chickens, ducks, and cows, the peacocks were part of the baggage when we moved our farm operation sixty miles from Franklin to Normandy. I had a nice big coop, so transporting the fowl would be no problem. Using my wire loop chicken catcher and chasing the slow waddling ducks on land made both chicken and duck captures easy. Climbing the tree at night to capture the peacocks wasn't an option. They needed to be caught during the day in a confined space. Unlike Joe Bowers, I didn't have an enclosed stall or feed room to lure them into. I did have a chain link dog pen that our dogs never used. I added chicken wire netting to the top of the pen to make what I thought would be a perfect trap. I sprinkled some corn outside and inside the pen and left the door open. I watched for a few days as I went about my chores, but never saw a peacock in the pen. The corn was gone, so I figured they must have been there sometime.

One day I replenished the corn and herded the eight peacocks in the direction of the pen. They were so wary of what I was up to, they took no notice of the corn or open door. As I crowded them closer to the door, some went right and others left, but not even one ventured inside. I was going to have to find another way to outsmart these wily birds.

I used some of the chicken wire and created a path that would funnel them toward the door. The next day I resumed my herding. Does one herd peacocks? It sounds better than flocking. Herding or flocking, it wasn't successful. I guess they recognized it as a trap because about the time they got to the open door, they would turn around and dash past me. I knew they had been in the pen several times since they were going in and eating the corn, when I wasn't around.

The next day's plan would involve more stealth. I would tie a string to the door and run it along the ground to a spot where I could hide and watch. When they went inside, I would pull the string, close the door, and capture them at last. It worked. Mostly. The next day I replenished the corn, flocked them to the vicinity of the pen, and took up my hiding place behind a car. I could see their legs by looking under the car. One by one, as the corn outside diminished; they would warily step through the doorway where the corn was abundant. In just a few minutes I saw five of the eight were inside the pen, but the final three were just outside the door. They would peck one grain of corn, raise their heads, and look all around to spot the man who herded them like cattle. The first few minutes of lying behind the car watching the peacocks slowly pecking, walking, and watching was rather peaceful. But after a while, the rocks of the driveway began to be an uncomfortable bed. How long would I have to lay in wait? How long would the corn inside the pen last? The first corn to disappear was that outside the pen. The three peacocks outside the pen looked first inside the pen and then surveyed the landscape. They were nearing a moment of decision, and I needed to be ready to pull on the string to spring the trap while all eight were in the pen. They decided the risk wasn't worth a few more pecks of corn and began stepping away from the door. I decided five peacocks was better than none and pulled the string to close the door. It worked perfectly.

The three sentry peacocks were startled and scurried off. It became an easy job to carefully catch the five peacocks by their skinny, brittle legs and put them into the coop. It was an uneventful drive in my pickup truck with the coop in the back. The peacocks were quite a curiosity at the country market where I stopped for gasoline and in the line of small-town traffic as I waited for the traffic lights to change.

When I arrived at the new farm, I decided it would be best for the peacocks to spend a few days in the chicken house to get acclimated to their new home. I unloaded the coop and placed it in front of the open chicken house door. When I opened the coop, they calmly, one by one, stepped out of the coop and into the chicken house. I checked on them every day for several days and always found them quite calm. I decided it would be best to release them from the chicken house late one afternoon. The few hours of daylight would be enough time for them to explore their surroundings, not wander too far, and find a good, high tree limb for roosting. There were several tall oak trees near the chicken house, so I figured one of them would be their choice. I didn't want a repeat of the Craigs' experience with their new peacocks.

The coop was still outside the chicken house door, so when I opened the door and moved away to let the peacocks find their way out in their own time, one by one they hopped onto the top of the coop. When all five were on the coop, they scanned the surroundings for a few seconds, and then in one startling, simultaneous leap, all five jumped into the air and sailed down the hill. They landed about fifty yards from the coop and walked into the woods. It was the last I saw or heard of those picky peacocks. My neighbors, the McBees, lived about a mile from our house, and they reported seeing them for a few days around their farm, but the McBees must have been like me, short on luck, because the peacocks chose to go somewhere else to live.

I acquired my next pair of peacocks as unexpectedly as I did my first. When our neighbor and mail carrier, Mildred Hall, died, her family asked me to officiate the funeral. I always refuse payment when I do a funeral for friends or neighbors, but Mildred's daughter, Glenda, kept saying she wanted to do something for me. One day I was at her mom's house helping her clear away some of the household items. There were as many peacocks roaming around Mildred's place as there were at Joe Bower's farm. Mildred even had some white albino peacocks. When I complimented how pretty they were and told Glenda the story of how I used to have peacocks, she said I could have a pair as a thank you gift for doing her mom's funeral. I agreed to come back some evening and get a pair. Several years have passed and I still haven't claimed my peacocks. Maybe I'm just afraid of testing my luck.

Twins and Their Aunt

My cows have had more than their share of twins. It usually happens about one in every thousand births in cattle. I had already had four sets of twins in my herd of twenty-five in twenty years, so the experience wasn't new to me. When I found Miss Centennial 23C standing in the field with two wobbly calves looking for their first meal, I felt like I had received a bonus. At the same time, I knew there were problems to overcome. I just didn't realize at the time that I was in for a twin challenge I hadn't seen before.

Part of the fun of twins is seeing double every time they do something. I'm not sure the mother thinks it's fun. It takes her awhile to realize there are two instead of one. It's not uncommon for her to give birth to the first one, get up and walk a distance, lie down and give birth to the second. She may or may not remember the first one. I nearly ran over one of my first twins with a truck. I had driven across the pasture to check on the cows when I found one with a new calf. After inspecting, weighing, and tagging the calf, I got into the truck and backed up to drive away when I caught a glimpse of something in the high grass through my rear-view mirror and hit my brakes. "Is that a calf? Who does it belong to?" I stood five feet from the truck tires, looking first at the newborn and then to the cow and calf I had just left, and then back at the little guy hiding in the high grass. I picked him up and carried him over to his mom and brother. She licked him on the head as if to say, "Oh yeah, I remember you. Where have you been?"

Some mothers of twins are so pre-occupied with cleaning and nursing the second one, they don't go back and take care of the first twin. That wasn't the case today with Miss Centennial 23C. She had both of her twins at her side. Both were heifers: one tan like her mother and the other light gray like her daddy. The tan first born was more active in her search for mom's udder, while the gray second born was getting her final cleaning up from mom. It was at this point that the first of several complications began.

When cows calve, it's not uncommon for other cows to come around to inspect the new arrival. Two or three cows arrived about the same time I did to join the welcoming committee. I'm sure all this attention was something the new mother could have done without. Giving birth was bad enough. Now I'm trying to keep my eyes on not one, but two newborns and all of you show up! One of the welcoming committee, number 016J and now called The Aunt, made her quick inspection and started walking away from the milling crowd. The tan first born started following her. Like most newborns, calves have blurred vision for a few days and will follow almost anything big that moves. I've even had them follow me when I finished my inspection and headed for the house. I just stop and turn them around, sending them back to their momma. When I saw the tan twin following this cow, I knew if I didn't correct this right away, it could be trouble. So, I herded The Aunt and the wayward twin back to where the welcoming committee encircled the momma and other twin. I slipped between the The Aunt and twin to chase her and the other cows away, leaving the new momma with her twins. The welcoming committee, including The Aunt, gradually dispersed, but I watched for a few minutes to make sure they didn't return. As the autumn sun began to set, I made my way to the house wondering what I would find tomorrow.

My first chore the next morning was to check on the twins. My anxiety was confirmed quickly. Standing across the creek from the birth site was the mother cow and the gray second-born twin. A quick scan of the close-cropped autumn pasture concluded with the tan twin nowhere in sight. Did she walk away and leave it? I crossed the creek to the birth site, but it wasn't there either. Had she tucked it into the high weeds along the fence row for safe keeping? Was it in the creek? Had it stumbled into the pond? I walked past all the places I thought the twin might be. This type of search wasn't unusual, even in normal single births. I often find cows who have obviously calved in the past day, but who have hidden the calf so well I can't find it. Following her isn't always helpful because she could just as well walk away from the calf to lead me away from it as she could walk to it. After checking all the likely spots with no success, I made my way back to the cow and twin. Looking out across the pasture, I was surprised to see two shapes coming my way. It was The Aunt with the tan twin trotting along behind! The sight was a mixture of surprise, relief, and concern. I was surprised to see the two of them together. Even after the episode of the afternoon before, I hadn't considered the possibility of The Aunt returning to steal a twin. I was relieved to see the twin alive and with enough vitality to trot across the pasture. She must have gotten some nourishment from somewhere.

My concerns came in bunches. Had the twin gotten the vital colostrum milk from her mother in the first few hours? It contains the antibodies and other nutrients essential for a newborn's start in life. Was the twin nursing The Aunt? She wasn't nursing a calf of her own, but was a few weeks away from calving and her udder was beginning to develop. Had the twin nursed at all from anyone? Would her mother take her back? If she didn't, was there enough milk in The Aunt's udder to sustain a newborn. What will happen in few weeks when The Aunt calved? One cow would raise two calves. Which would it be?

One question was answered right away. The Aunt neared where I was standing with the mother cow and calf. When she stopped walking, the trailing calf trotted right to her side and put its mouth right into her udder. The calf was at least nursing someone. Only time would answer if The Aunt had enough milk to sustain the newborn. I still thought the best arrangement was to return the twin to her mother as nature intended, so I repeated my herding-switching routine. I brought the two cows and two calves together easy enough. The wayward twin even nuzzled her mom's udder a bit, but when I separated The Aunt from the group, the twin scampered after her. The abduction was complete! Nature had taken a new course. I made my way across the pasture mulling over the unresolved issues. By the time I arrived at the house, I had concluded there was nothing else I could do but watch and wait.

My daily check found both twins doing fine. The adopted twin seemed healthy and was getting enough nourishment to grow normally. This arrangement might even be the best arrangement for now. Both twins could grow to their full potential. Even though weaning two calves of four hundred pounds each instead of the normal one five-hundred-pound calf resulted in more overall production, twin heifers seldom are kept for replacements to the herd since they would always be smaller. This arrangement might let me keep two heifers from one of my best cows. But those ideas always ended with the question of "what is going to happen when The Aunt gives birth to her own calf?" Nothing to do but wait.

The wait lasted four weeks. Late one afternoon I was driving my truck on the farm driveway between our house and the inn at Parish Patch, when I saw a cow stretched out under a towering oak tree. Even in calving season, one's heart skips a beat when one sees a cow alone and stretched out. It doesn't mean they are taking a nap and the rest of the herd has slipped away. She is sick, calving, or dead. During calving season, one hopes she is in labor and everything is progressing normally. It's better to stay away and let her calve on her own. Someone's presence just makes her nervous and interrupts a normal process. Interruption at the wrong time leads to intervention. She stops pushing with the less frequent and lighter contractions and one has to pull the calf without her help.

I drove the truck within thirty yards of where the cow was lying and parked quietly. Our golden retriever, Lane, jumped out of the back of the truck and was on her way to investigate. This is not good. If my presence makes a calving cow nervous, a dog makes her berserk. I called for Lane to come back, and when I did, the cow raised her head. It was The Aunt. As I moved around the truck to put Lane in the cab, I noticed the adopted twin lying by her side. The aunt was well along in the process, but at a critical stage. I could see two feet and a nose protruding. That was good. The calf was in position for a normal birth. Its head, often too big to comfortably emerge, was already past the pelvic circle. It's at this point calves often get stuck and need help. But just arriving and seeing these indications doesn't tell someone whether it's a normal birth and they just arrived at this stage or if she's been stuck here for hours with no progress. Nothing to do but wait.

Often, I will leave for a half hour or so and come back to see if there is any progress. If there has been, I leave and let her finish the job on her own. If there hasn't, it's time to give some assistance. When the cow seems undisturbed by my presence, I move away a distance and out of her vision, but close enough that I can monitor the progress. During calving season, I keep binoculars in the truck so I can have a close view from an even greater distance.

Lane knows something special is happening as she sits on the seat beside me watching alertly. She looks at the cow and then at me, as if to ask, "What are you going to do next?" As she looks back at the cow, my answer is, "Watch and wait."

As we watch, something happens I had never seen before and probably never will again. The twin gets up and stands by "his mother's" head. Their noses met. "What's going on mom?" the calf seems to ask. "I can't explain it right now," she answers. The twin then walked around her and began nursing from her udder! Great! She's trying to give birth and this calf wants dinner! I know mothers often have to do several things at once to take care of their children, but this seemed to be expecting entirely too much.

I wasn't sure whether to just continue my watching or risk disturbing the cow and break up this sordid scene. Before I could decide, the situation got worse. The first member of The Welcoming Committee arrived in the form of whom else, the mother of the twin! Was she here to take back her twin? Frankly, the Aunt would have probably welcomed the idea at the moment. Go suck on your own mother! Or was I about to view my rare, but second, abduction in a month. Even though I had only been waiting ten minutes, I couldn't just watch any longer. I had to break up this sordid situation. First, I chase away the mother of the twins and then I moved the adopted twin away from The Aunt's udder. For a moment I thought The Aunt was going to stand up, but as I moved back to the truck, she laid down in response to another contraction.

Lane and I watched. The adopted twin watched. The mother of the twins and her single calf watched. The Aunt strained. But there had been no progress since I arrived. All I could see of the new calf was two front feet, a nose, and a mouth with a long thick tongue hanging out the side like the tongue of a panting dog in the heat of summer.

The time of waiting was over. She needed help. It was time for action. Timing my approach with the frequency of her contractions, I slipped quietly behind her and out of her line of vision. When the next contraction came, I grabbed the feet of the unborn calf and pulled with all my strength. In some circumstances such a pull is like pulling on an immovable object, but this time there was a small, gradual movement. First the eyes appeared and then the ears. Adjusting my grip on the calf's slippery ankles, I gave another pull, and those big shoulders broke through the pelvic opening with a rush. We were home free. The eyes opened. The tongue moved. The head shook, as if to shake off the memory of a bad experience. We, for sure, had a live one here. The rest of the chores would be up to momma. She had a new baby bull calf. As I pulled the calf around to the cow's head, The Aunt raised her head and began cleaning her newborn. The licking a new calf gets from head to hoof, not only gets it cleaned up, but stimulates circulation and movement like a giant massage. The next hurdle for The Aunt was standing up. In difficult births cows will strain muscles, ligaments, and joints so much that they won't or can't stand up. The longer they lay there, the more likely they won't ever get up. Having been through that with more than one cow, I knew getting her to stand in the next few minutes was critical. I slapped her on the rump and she stood up right away.

Right on schedule, The Welcoming Committee returned. The twin's mother took over the licking process. This wasn't unusual, but given the adoptive history of these cows, I wasn't enthused about the prospects. Before I could decide if I should do anything, The Aunt walked a few yards away, obviously still having contractions. This was not the behavior of a protective mother. Once a cow is on her feet, some won't let anything or anybody near the new calf. But right now, I was watching the adopted twin trying to nurse The Aunt on one side and the mother of the twins licking The Aunt's newborn on the other.

It was time for action again. The Aunt had to bond with her new calf. The new calf had to get on its feet and get its first colostrum milk. But wait. The Aunt didn't have any colostrum! She had been nursing an adopted twin for four weeks. Regardless, the new calf needed to nurse in the next few hours. First, I had to scatter The Welcoming Committee. Then I had to get The Aunt back to her new calf. I lifted the new calf to his feet. He swayed back and forth like a drunken sailor on a stormy sea. After a minute or so, he decided walking was a way to get some control and took his first step. He stepped too far and leaned back to correct himself only to over-correct. With a lurch forward, he ended up on his face in a pile of flailing legs. I helped him gather himself and put him on his feet again. This time he wobbled off in the direction of his mother, looking for that first meal. After an unsuccessful search between her front legs, he found her udder on the other end. I don't know if his swollen tongue was a problem or not, but he couldn't seem to get a teat into his mouth. After watching him staggering from one end of the cow to the other for a few minutes and never nursing, I decided I should just leave them alone for a few hours and see if they could manage on their own. So, I headed for the house and left The Aunt with her adopted twin and newborn calf all standing alone under the oak tree.

After supper I went back to check on them. All three were lying close together still under the oak tree. The Aunt looked fine and so did the newborn. I assumed he had nursed, even though I had never seen it. Waiting until tomorrow would help me know.

The next morning it didn't take long for me to find the newborn. He was laying a few yards from the pasture gate. The problem was The Aunt and the adopted twin were standing a hundred yards away. Had they left him or did she put him there for safe keeping? He looked alert, and when I walked over to him, he stood up and sprinted off toward The Aunt. That looked good. At least he knew who he belonged to. Had he nursed? With that energy, probably yes, but I couldn't be comfortably certain unless I got to see it. When he arrived at his mother's side, he went straight for her udder. At least he had figured out where it was located since last night. His tongue wasn't hanging out like yesterday. As he started to nurse, she looked around at him, raised her leg, and stomped it back down. It was like a slap in the face for a suitor who had touched a lady where she didn't want to be touched. Undaunted, the calf bounced back and tried again only to get the same treatment. This time he walked around to the other side, hoping the leg on this side would stay still. Her response this time was to turn her head and butt him away. One didn't have to be an experienced cattleman to know this wasn't working and waiting wasn't going to help.

Not all cows are natural mothers. Some walk away after calving and abandon their calves. Others won't let them nurse. It may be that they identify the calf with the pain of birth and decide not to have anything to do with it. One remedy is to put the cow and calf in a confined area, like a stall in the barn, for a few days so she can't walk away and leave the calf. Cows that won't let their calf nurse can be restrained in a chute where the calf can get to her and nurse without being kicked or butted. A day or two of this and both can be turned back to the herd without problems. I decided taking these three to the barn was my best strategy. We were at least a quarter mile from the barn and, while I could herd The Aunt and the adopted twin, I didn't think the new calf could keep up. I wanted to try anyway. After fifty yards it was obviously not going to work. The newborn kept falling further behind as the others kept moving toward the barn. I herded the two on to the barn with the thought of coming back for the new calf to herd him by himself. When I got back from the barn and started him toward the barn, I realized with the short steps he was taking, this trip could take all day. Normally I would put him in the bed of the truck and haul him to the barn, but my wife, Claudia, had driven it to Nashville that day and left me with our Lincoln Town Car. I could use it. After all, it had a huge trunk and the calf could stand up in it for the trip to the barn. It worked, but I'm glad no one drove up behind me as I eased down the driveway to the sound of a puzzled mooing calf in my trunk.

After settling them into the barn, I watched to see if there was any change in The Aunt's willingness to let her calf nurse, but I didn't see any. It was time to wait again, so I left them to work it out for themselves and went about my work for the day. Late that afternoon, when I checked on them again, there was still no indication that The Aunt was giving any attention to her new calf. It was time to try something different while the new calf was strong. I would separate the two calves from the cow overnight and the next morning would turn the cow back to the calves, hoping that their eagerness to nurse and her desire to be nursed would make her willing to let the new calf nurse. It was time to wait again.

The next morning both the cow and the calves were eager to get together. I opened the gate, and the cow went right into the stall to stand between the two calves. The adopted twin went straight for the udder on one side and her new calf nosed in from the other side. The cow put her nose into a bucket of sweet feed I had put down for her. I was hoping all the activity of this three-ring circus would distract the cow enough to let the new calf nurse. With feed dribbling from her mouth, the cow looked back, first left then right, to evaluate the situation. Her decision came quickly. One of these calves was an intruder. She gave a light, but sudden, kick to the new calf that spun him around. I didn't need to see any more. I moved into the stall and herded the cow and her adopted twin out the door. The undaunted newborn wanted to follow, but I blocked his way. As I turned to close the door, he stood by the partially eaten bucket of sweet feed with a confused look. "Hey, that's my mother going there," he seemed to say. "I belong with her. Are you confused?"

If I was confused, I wasn't alone. The Aunt, the adopted twin, the mother of the twins, and this new calf were all confused. I deserved to be confused too? As I closed the stall door and left the little guy standing there alone, there was one thing I wasn't confused about. Who would be this calf's momma? It would have to be me!

A New Buddy

Raising baby calves on milk replacer and a nippled bucket isn't uncommon. In fact, all dairy calves are removed from cows after three days when all of the early colostrum milk is gone. It's not as common in beef cows, but is done when the mama dies or won't accept her calf. That's what I faced when the Aunt, who had adopted one of the new-born twins from another cow wouldn't accept her own calf, when it was born a month later. After becoming certain The Aunt wasn't going to accept her new calf, I turned her and her adopted twin back into the pasture.

My mind was back at the barn with the now orphaned calf. Had it nursed even once since it was born twenty-four hours ago? Surely it had, but I hadn't seen it. Only its strength and mobility suggested nourishment had come from somewhere. I knew it didn't get any colostrum. That was strike one against it. If it hadn't nursed at all, that was strike two. He could be in trouble.

On occasion I keep some colostrum in our freezer at home. It's helpful in emergencies like this. I get it by milking cows whose calves died at birth. But I didn't have any in the freezer now. I did have two other things I kept on hand for similar situations. Some calves are born weak and need a stimulant or supplement. One of the old home remedies granddad used on calves and himself, as a matter of fact, was a shot of whiskey. It was put in with the calf's milk, but granddad skipped the milk. I've used this remedy on calves when there weren't other alternatives, but like with people, it has limited short-term benefits. My preference was to keep a package of electrolytes, sugar, and vitamins on hand. It's a high-powered Gatorade for cows. I also kept a supply of powdered milk replacer for calves.

On my way back from the pasture, I stopped by the shop to mix up a two-course dinner for the new calf. With him still being eager to nurse, there should be no problem getting these down him in a few minutes. I dissolved each solution with warm water, testing the temperature like a mother mixing baby formula.

For the first course, I put the electrolyte solution in a plastic half gallon bottle and screwed on the rubber nipple top. It was only half full, but when he also got a quart of milk replacer it would make a full meal. It may even be too much for one feeding. I would let him decide when to stop nursing on the milk replacer. He would know when his stomach was full. He needed all of the electrolytes and would get them first. I put the powdered milk replacer into a galvanized bucket that also had a rubber nipple. It was used as the calves got older and ate more.

When I opened the stall door, I found the orphan lying in the spot where he stood when I saw him last. Whatever the reason, tired, sad, or weak, he decided to just plop down right there. I picked him up the way cows get up from laying down, hind legs up first and on his front knees, and then up on all fours. He just stood there, neither offering resistance nor escape. I put the nipple of the electrolytes bottle to his nose, expecting him to recognize it as a close resemblance to something nature intended and begin nursing. His head dropped in disinterest. I lifted his head and poked the nipple at his mouth. It didn't open. I slipped a finger into the corner of his mouth and, when it opened, I inserted the nipple. Nothing moved. There was no nursing. It was as if I had put a stick into his mouth. Maybe he knew it wasn't the real thing and didn't know what to do. I reached inside his mouth and squeezed the nipple between my thumb and finger, feeling the warm electrolytes ooze out of the nipple. He may have swallowed, but he still didn't nurse. I squeezed his upper and lower jaws together several times with my hand, trying to mimic the nursing action. There was no response, except for a drool of electrolytes coming from the corners of his mouth. He wasn't even swallowing.

Was he beginning to shut down? Was there a problem with the tongue that was swollen yesterday? Would he swallow? Could he swallow? The electrolyte sugar mixture has the consistency of molasses, so maybe it was too thick. I could try the milk replacer and hope for a better result with the planned second course. I repeated each step. Present, poke, insert, squeeze, and press. The responses were all the same, except at the end, milk replacer streamed out of the corners of his mouth and down his chin, instead of the drool of the electrolytes.

I had one more trick to try before resorting to desperate means. Orphan calves being raised on a bottle will suck on about anything: ropes, sacks, other calves' tails, but especially your fingers. They are just about the right size as the real thing and will serve as a pacifier, even if they don't give milk. Some farmers never start calves nursing from a nipple on a bottle or bucket. They teach them to drink from day one by making them think they are nursing. They get the calf started sucking on their fingers and then ease their hand down into a bucket of warm milk replacer. After the calf starts sucking milk up and around the fingers, they slowly withdraw their fingers, and the calf usually continues to drink. Some calves require a few repetitions before they learn to drink without the fingers, but it eventually works.

I set the bucket of milk replacer on the stall floor and slipped two middle fingers into the calf's mouth. He wouldn't suck on them. Only after wiggling my fingers inside his mouth did he offer a response and then it was just an unenthusiastic chewing, tongue wallowing motion as if to play with my fingers. Regardless, I moved to step two and put my hand in his mouth and down into the bucket. He didn't want to put his head down and resisted my efforts to push it down with my other hand. I lifted the bucket to his face until his mouth and nose were in the milk. Nothing happened. Even the wiggling of my fingers got no response. In frustration, I put his face into the milk half way to his eyes. By now I would have celebrated if he had just blown bubbles from his nose into the milk. He just shook his head free from the bucket and backed away.

I had to get something into his stomach or he would be a goner by morning. Fortunately, I was equipped for more drastic measures. Some young calves get so sick, they stop nursing or eating. They have to be fed by running a tube down their throat and attaching a bag or bottle of fluids that can flow directly into the stomach. It's called drenching and is also used for giving certain medications to mature cows.

This orphan was going to get his first meal straight into the stomach. I took my bucket and bottle back to the shop and transferred the electrolytes into the drenching bag. There would be nothing natural about this approach, but it may be the only thing that would keep him alive. I threw away the milk replacer. I didn't want to risk overfeeding him or, even worse, forcing so much liquid down him that he drowned. Several small feedings, four hours apart, were more effective than two large feedings a day. I would start with just electrolytes now and add milk replacer later.

Back in the barn, I found him curled up in a corner of the stall with his head propped up by his chin resting on the floor. He was looking worse to me as the hours passed. At this rate, he may not live through the day. I lifted him to his feet again. Standing straddling his back, I held the drenching bag in one hand and tried to insert the tube into his mouth with the other. I needed a third hand. As I poked the tube at his clenched teeth, he shook his head and started backing from between my legs. I squeezed my legs together to try and hold him, but he slipped away. Nothing about this process interested him. I backed him into a corner of the stall, so the walls could give a helping hand. Straddling his back again, I held the tube with part of my hand and used some fingers to pry open his mouth. It was then an easy matter to gently slide the tube along one side of his mouth and down his throat. The tube had an enlarged bulb on the end to keep it from accidentally going down the windpipe. Before this feature was added, calves would drown or get pneumonia from fluid in their lungs.

Since I didn't have to depend on his swallowing, the electrolytes drained out of the drenching bag in a matter of seconds. I withdrew the tube, and he walked away shaking his head. That was a rough way to get his first meal, but at least I knew it was in him.

By now it was ten o'clock in the morning. I would return in four hours with another dose of electrolytes in hope of finding a more energetic calf or at the very least a stable one. It was time to wait again.

At lunch time, I called our daughter, Whitney, who was away at Vanderbilt. After catching up on her news, I gave her the latest chapter of the adopted twins' story. She wanted to know if I had named the orphan yet. When I told her I hadn't, she suggested Buddy. "Looks like you're going to have a new buddy," she said.

Not every farm animal gets a name. Some cows are stuck with a tag number that corresponds to their birth date. Buddy could have been known as O16 since he was born on October 16. Others get named for friends or family who are visiting when a calf is born. Some unusual event or behavior may be the source of a name like The Aunt or Orphan Annie. When we put a cast on a calf's broken leg, he became known as Tripod.

The mid-afternoon "feeding" of Buddy found him in the same condition as at mid-morning. He was lying in the straw looking neither better nor worse. In hopes of getting him to nurse or drink, instead of using the drench tube, I mixed some milk replacer and put it in a nipple bucket. I repeated all of my tricks with the nipple and had no more success than I had earlier. So, after I drenched him with the electrolytes, I filled the bag with milk replacer and let it flow into his stomach.

The sun had already set when I returned after supper for his final feeding of the day. I opened the stall door, turned on the light and saw Buddy curled up in the center of the stall. He turned his head toward me and blinked his eyes several times to adjust to the sudden light. In optimism I had mixed a quart of milk replacer in the nipple bucket. I wanted to keep trying to get him to nurse. My cautious side made me bring along the drench bag and tube just in case. I lifted Buddy to his feet and presented the nipple to him. He had no recognition or interest in it. Even when I forced it into his mouth, he just let it rest there or chewed on it when I wiggled it. Once again, he missed the taste of a meal as I resorted to the drench tube.

The next morning Buddy was standing in the stall waiting for me. This was my first sign of turning a corner. Maybe he would be interested in nursing. I would try and be more persistent. In this case, persistence didn't pay. Buddy used his new energy to fight the nipple instead of looking for it. I had run out of tricks for making him nurse or drink. How long can one continue drenching a calf? He would get a quart of milk replacer each morning and evening until it was gradually increased to two quarts each feeding. Would his throat get irritated from the tube sliding in and out twice a day? As the days and weeks passed and Buddy grew stronger, how easy would it be to get the tube down his throat? This was new territory for me, and I didn't have any answers to my questions. I would just have to wait and find out.

For the next three days I didn't bother to take the nipple bucket. Buddy got his feedings from the drench tube and seemed to grow to expect the routine as if this was how all calves did it. The drenching was over in about a minute where it would have taken him several minutes to nurse or drink from a bucket. But that was the only advantage to the situation. I was looking forward to seeing my neighbors at church tomorrow. One of them might have a trick I had never heard before.

It's common for my Sunday school class to begin with conversations in several little groups. As they sip coffee and nibble on homemade goodies, they catch up with each other on the events of the week. The conversations range from national to local and from personal to abstract. When there was a simultaneous break in the conversations, I seized the moment to tell my Buddy story and solicit their advice. I hadn't told the adopted twin part of the story, so I built up my dilemma step by step.

When I posed the problem of a calf that wouldn't nurse, there was ready support from several around the room. They had been there and knew the problem well. Unfortunately, all the solutions offered were the tricks I had used in failure. Even Jim Cortner, the octogenarian who had been a dairyman all his life, until he sold his herd the year before, wasn't able to help. "You may as well knock him in the head, that calf will be more trouble than he's worth," Jim said. I didn't know what I was going to do, but I knew I wasn't going to knock Buddy in the head. I would just keep drenching him until either he or I figured something out.

Monday morning, I went to the farmer's co-op and bought a sack of calf starter. With bucket fed calves the goal is to get them off of milk replacer in about four to six weeks and onto dry feed. It's a gradual process and the hardest part is getting them to begin eating any dry feed. Calf starter is laced with molasses to make it more palatable and encourage them to begin eating it. Given my experience with Buddy, so far, I would need every advantage I could get. If I could get him on a full ration of calf starter sooner than normal, the drenching/nursing issue would be resolved.

That evening I gave Buddy the normal drenching and then put a few grains of the calf starter into his mouth. He started chewing on them right away like a kid with his first candy bar. He made quick work of it, so I gave him some more. For once, he was acting normal. I put something in his mouth and he knew what to do with it. I put a cup full of the calf starter in his feed bucket and lifted it to his nose and mouth. He didn't offer to eat any out of the bucket, but I was content with his initial interest and set the bucket down against his water bucket. He would know where to find it if he decided it interested him later.

The next morning, I was surprised to find all of the feed gone from his bucket and even the water level lower in the water bucket. I checked around the feed bucket to see if he had turned the bucket over and spilled the feed, but there was none on the ground. He must have eaten it! This was good news and, if he had also drunk some water, it was a great day. This situation was turning around. Both were small steps, but they held promise for a good resolution to what had been a series of difficulties so far. Now all I needed to resolve was the twice daily drenching. Buddy needed to learn to nurse or drink the milk replacer or face drenching until he was a month old.

The rest of the week was uneventful. The twice daily feedings became a routine. Every morning and night Buddy was drenched with a quart of milk replacer and got a cup of calf starter. During the weekdays there was little time to do more than what the feeding chores required, but on Saturday morning there was time to loiter and enjoy the wonder of a newborn. After the drenching, I sat on a bucket and watched Buddy explore the sweet-smelling calf starter. He seemed content just knowing it was there to eat later. I got some out of the bucket and put it into his mouth. Before I could get my fingers out, he started sucking on them. So, he did know how to nurse! I pulled my hand away, but he followed it as he chewed and swallowed the bits of grain. His head pursued my hand like a fighter plane locked onto its target. He wanted my fingers again.

This wasn't the first time a calf had nursed my fingers. When I was in high school, I raised a couple of Holstein calves on the bucket. After they had emptied their bucket, they searched for a replacement for the nipple that had been removed. Almost anything would seem to do at the moment. Touching one of them on the nose, he grabbed my fingers and resumed the nursing as if he expected another quart of milk out of my fingers. All was well until the calf became dissatisfied with no results from his nursing efforts. His instinct took over and he butted my hand like he was butting his mother's udder when the milk had run out. I don't know what it accomplishes, but it's a natural behavior for calves. I wasn't expecting the butt when it came and was surprised when a sharp pain hit my fingers. The upward motion of the calf's butting had thrust his lower front teeth into my two middle fingers between the knuckles. Cows only have teeth on the bottom in the front, but this was enough for this calf to give me wounds that made scars I still carry today.

I was ready for Buddy's butt and let my hand flow with his motion. After the butting, he renewed the vigor of his nursing, as if expecting the action had started the milk to flow. Buddy was becoming more vigorous and insistent with his nursing and butting, so I pulled my fingers out of his mouth and hid my hand behind my back. He wasn't about to give up so easily though and started searching for my hand the way he searched for his mother's udder the first hour he was born. He poked at me all over and finally found a snug resting place under my arm for his head. It must have resembled reaching under his mother's leg to find her udder. There was no udder under my arm, but he did find the sleeve of my coat to be a satisfactory substitute at the moment. How could I use this instinctual search for an udder to get Buddy to nurse the nipple of a bucket? For now, I had to standup and get away from him before he ate the jacket off me. I stood up and turned away from him to walk away, but he quickly followed. Before I could get away, he pushed his head between my legs and gave a butt. It could have been painful, but I reacted in time to protect myself. He definitely had found the nursing instinct again. I would return for the evening feeding prepared to take advantage of his instincts.

I told Claudia about the events of the morning feeding and of my idea for capitalizing on Buddy's instincts to get him to nurse the bucket. I would try a trick I had never seen or heard of. That evening Claudia went with me to the barn to feed Buddy. When I opened the stall door, he romped in my direction eager for his feeding. This time I had brought the nipple bucket instead of the drench bag. He was so eager; he nearly knocked the bucket out of my hand. Maybe I could just present the nipple and he would go for it. I put the nipple in front of his nose and he circled it with his mouth like a blind man searching for a door knob. His mouth searched all around the bucket, but even when he touched the nipple, there was no awareness of what to do with it.

It was time for the trick. I turned around and moved to the center of the stall with Buddy in hot pursuit. Just like at the morning feeding, he pushed his head between my legs from behind. When his nose appeared out the front side, I put the nipple in front of him and he grabbed it like it belonged to his mother. As the nursing began, I looked at Claudia and smiled a smile that said, "I know this looks kinky, but it's working."

After a minute of concentrated, uninterrupted nursing I decided it was safe to reveal the trick to Buddy and stepped over his back holding the bucket in front of him. He never missed a beat. After Claudia congratulated me, she asked, "Are you going to report this to the Sunday school class? How will you describe it?" I wasn't sure how to tell the story, but I knew it would raise some eyebrows!

The next morning, I didn't have to conceal the bucket nipple. I just put it in front of Buddy and he nursed enthusiastically. After a minute or so of holding the bucket for him, I hung it on the stall wall and let him finish nursing while I scooped grain into his feed bucket and gave him fresh water.

I didn't have to wait for a quiet moment to break into the conversations with the Sunday school class. The first question was "How's Buddy? Have you got him to nurse yet?" I took them step by step through the events of the week from the eating first grain to the nursing on my coat sleeve. Then I introduced the thought of my new found trick and stood up in the middle of the class to demonstrate the between the legs from the rear disguise. Amid the hooting, hollering, and laughing, Jim Cortner, the local dairyman, said, "Well that's a new one on me and I thought I had seen it all!"

Born Mean

Some become mean because they have been abused or learn they have to be mean to survive in a tough world. Others just decide they want to be mean to stay on top of the pecking order. I think she was born mean.

We first met in October on a consulting trip to the prairie country of North Dakota. I was attracted to her on first sight. It would be another four months into our relationship before I would discover her mean streak. It would be years later before I decided to end our relationship because she was just born mean.

I was sitting comfortably in the seat of a pickup truck, when I first saw her. I never got out of the truck, but had a good view of her. In retrospect it might have been a good idea to have gotten out of the truck and walked right up to her. This was a case of distance lending enchantment to the view.

She was to be the third Beefalo cow I bought. Had she high tailed it over the nearest hill, I might have changed my mind on the spot, but she just stood there lazily chewing her cud and swishing her tail. I didn't even notice a mean streak a month later when I rendezvoused with the owner in Louisville at the North American Livestock Exposition for the delivery. We backed our cattle trailers up to each other, opened the gates and 298 stepped from one trailer to the other, like she was at home and moving from one pasture to another.

The name on her registration papers was Halls Miss Beefalo 298, but I just called her 298 which was the number on her ear tag. It was also permanently tattooed in one of her ears. Since cows were first domesticated, farmers have given names to their cows that were associated with a physical or temperamental characteristic. One with a white spot on the forehead would be named Star. Some names could be unflattering like Big Butt or Droopy Bag. I'm sure the famous Bossie earned her name. I've owned a Tripod, Praying Cow, Bad Eye, The Twins, Mister C, Big Red, Long John, and Taffy. Registered animals get high classed names like Mr. Centennial Gold, Geronimo, or Sir William's Pride.

The owner of 298 must have just used consecutive numbers. I used numbers to indicate their birth month and day and the livestock industry's international letter for the year. My first Beefalo calf was Miss Centennial 427T which was born on April 27, 1976. For their registration name, I added Mr. Centennial for the bulls and Miss Centennial for the heifers which identified them on ancestry charts as coming from my farm, Centennial Acres. I still may have named this North Dakota cow, Miss Mean 298.

Three months after I bought 298, she had a calf who became Mr. Centennial 215U. I walked through a light snow across a small pasture next to our house, when I found them. I carried an axe and was on my way to the pond to cut a hole in the ice for the cattle to get drinking water. In the middle of the pasture, I saw 298 standing with a newborn calf nearby. The mother kept her eyes fixed on me as I approached, but didn't move. The calf was just a few hours old and had already been licked clean by its mother. I would come back later with the scales to record its birth weight for registration records, but for now I would make sure it was healthy and see which sex it was. Mother cows usually just stand and watch the inspection and weighing and that's what 298 was doing from about six feet away. As I reached down to raise the calf's hind leg for a glimpse between its legs, I caught some movement out of the corner of my eye. I looked up in time to see 298 coming quickly my way with her head down and fire in her eyes. I dropped the calf's leg and backed away, but she kept coming. She bounded over her calf and was within a foot of my midsection when I instinctively hit her across the nose with the handle of my axe. I didn't want to hurt her, but didn't want her to hurt me either. If she kept coming, even the axe handle would not have provided me with enough protection from her twelve hundred pounds of fury. Fortunately, the blow to her nose was enough and she retreated with a snort to stand over her newborn. We both stood motionless and just looking at each other with our frosty breaths clouding the air around us. With a quick glance at the calf, I decided it weighed about seventy pounds. I wouldn't be returning with my scales for an actual weighing. I could figure out its sex as it grew older.

This incident wasn't enough for me to label 298 as a mean cow. She was just instinctively protecting her young and some mothers are more protective than others, especially with their newborns. I would just remember next year to be cautious when I approached her newborn. She might be just fine with it, but it would be worth my paying close attention.

By summer I had been in the pasture near 298 and her calf several times. It was always without incident, but I still maintained a respectful distance. As the pasture grass became thin, it was time to move them to another field, so I herded them across the field to the gate. They walked along calmly and through the gate. As I returned to close the gate, I heard some noise behind me. I turned my head just in time to see 298 butt me in the rear end and send me flying into the air. She had come back to intentionally give me a message. I flew as high as the gate before I started coming down to land flat on my back. Stunned and breathless, I looked straight up into the face of an angry 298. I felt her hot breath on my face and saw the fire in her eyes, as she stood over me trying to decide whether I deserved more punishment. At that moment I didn't know if I could move or not, but before I tried, she decided she was done and trotted off after her calf into the new pasture. That was a close one! She didn't have horns but she could have done plenty of damage with her head, not to mention what one blow from her hooves would have done to my head. That settled it. This cow was mean.

The year went on without incident mainly because I didn't get close enough to provoke her. My farm wasn't the wide-open prairie of North Dakota, but I gave 298 as much space as I could. I never turned my back on her. She raised a very good calf, but it was always wild. She either trained her calves or they inherited her temperament. Since cows are kept in the herd based on their ability to produce good calves, I decided I should keep 298, but be wary.

The next winter when I found her in the woods with a new calf, I approached slowly and with no plans for getting close enough to touch the calf, much less weigh it. I would just have to determine its sex from a distance and guess its weight again. I didn't get nearly as close as last year. When I was ten yards away, 298 charged after me on a dead run. One can't outrun a cow, especially an angry one. The only thing that saved me this time was a tree I ran behind just as 298 was about to butt me. Her pursuit wasn't over. She turned on a dime and was after me again. I ran around to the other side of the tree, but had to keep going. Around and around we went. Fortunately, she couldn't run in circles as fast as I could, and she finally gave up. She trotted back to her calf and gave me an unsatisfied look of warning. I decided this cow woke up every morning looking for a chance to run me down. I was determined not to give her the slightest chance, especially when she had a calf.

At hay feeding time she was given plenty of distance. When I moved the herd between pastures, I liked for her to be in the middle of the herd. The closest I had to get was at the barn during the annual worming and vaccination. One fall I had all the cows in the feeding shed. It was after weaning time, so there were no calves. I was moving the cows through the shed toward a working pen and holding chute, when 298 decided she would rather get to me than go toward the working pen. I didn't have an axe handle for protection or a tree to run around. It was just me and a mean cow in a shed with walls too high to jump over. I couldn't escape this time. All I could do was minimize the damage. Boards from the hay manger were above my head, but too close together to crawl through. I jumped up, grabbed a board, and started to pull myself up and out of her reach. I lifted my legs to miss her first pass, but couldn't continue to hold them up. I wasn't fast enough and couldn't lift myself high enough to get beyond her head butts. When my legs dropped, she butted them with a ferocity I hadn't seen before or since. I'm convinced she wouldn't have stopped until she killed me, if I had been on the ground. Fortunately, my body swung away as it absorbed her repeated butts. I hoped she would give up soon because I couldn't hang there much longer. She backed up a step and took aim at me again. She missed as I raised my legs a final time. When they came back down, I gave her a kick on the nose to end the standoff. This cow was dangerous. Great production or not, she needed to have a new home.

That spring I sold her to Charles Elrod with a warning label, "Caution: Born Mean."

Mooing at the Opry

When you come to Nashville, attending a showing of the Grand Ole Opry is a must. It makes no difference that you prefer real opera or classical music and hate the sound of country or bluegrass music. It's a cultural event. I've attended quite a few shows, but none had the sound like the one in 1975. It was loud and melodious to my ears. They were having a cattle auction on the Opry stage. Cows, bulls, and heifers came to center stage and stood in the spotlight where Roy Acuff, Porter Wagner, Little Jimmy Dickens, and Minnie Pearl would perform the next night. The American Beefalo Association was holding its first national sale and wanted a grand stage to display their new breed.

The spectacle started the night before at a prime rib banquet at Roger Miller's King of the Road Hotel. I got my first taste of Beefalo meat and my first glimpse of a live Beefalo. For those who know more about opera than cattle, a Beefalo is 3/8 Bison and 5/8 domestic cow. I won't bore you with how they get those fractions to work out. It's enough to know that the meat is lean, low fat, low in cholesterol, and tastes like beef.

The view I got at the banquet turned out to be a little too close for comfort when they led a two thousand-pound bull into the banquet hall. It came down the aisle beside my seat at the end of the table. I could feel his breath before he got to me and looked up at his big brown, glassy eyes larger than a silver dollar as his head passed a foot away from me. If they wanted to impress me with how big this guy was, it was working. I felt small all of a sudden. I was tempted to reach out and let my hand flow along his fluffy furry coast as his massive side passed by blocking my view of the rest of the world. It's a good thing I didn't touch him, because those thoughts were a split second before a near disaster began to unfold. The bull's handlers had rolled a plastic sheet down the aisle to protect the floor from, let's say without being too graphic, an accident. But their protective measure nearly caused an accident. Beside my seat the floor changed from carpet to a wood parquet dance floor under the plastic. The bull's hooves had a good grip walking on the carpet, but his first hoof hit that wood floor and the plastic made it like an ice rink. Before he could regain his balance the second hoof slipped and then he had both legs clambering for some solid footing. He began to lose his balance as his head, legs, and body were all going in different directions. Unfortunately, that giant fluffy furry wall of his side started coming my way. Before I could get up out of my chair, he weaved in the other direction and landed on his side in the middle of the aisle with his feet nearly in my lap. I was hoping I wasn't about to get kicked as he struggled to get up. There was as much clattering as he got up as there had been when he went down. The bull was calmer than most of the guests. Because most of them were cattlemen, they knew what could happen with an angry bull in a crowded ballroom. I have one piece of advice if one is thinking of taking a bull into a ballroom. Don't!

My decision to attend the festivities started with wanting to get into the cattle business. Having grown up raising cattle on our farm and having a degree in animal science didn't make me a novice to the cattle business. I wasn't sure what kind of cows I would have, but knew an ordinary $1,000 cow ate the same amount as a $10,000 registered cow and her calves would sell for a lot more. Since Beefalo was a new breed and something different, I thought they might work for me.

Back stage at the Opry had never been like this before or since. From my seat at the foot of the stage, I could hear the cattle mooing backstage like they were warming up for the show. The curtains parted enough for the first animal to be sold to make its entrance. The spotlight hit its furry coat and made it sparkle. Its owners had been working for hours washing and blow drying. Dolly Parton never had as much time spent on her hair as this heifer had on hers. This was heaven for the auctioneer, Jerry Cash, (no relation to Johnny!) as the perfect acoustics of the Opry House made his melodic chant sound as good as the Statler Brothers. Of course, the Statlers never had such mooing for backup.

It must have all worked because by the time the auctioneer's gavel came down, the bid had reached $125,000. I was glad I hadn't scratched my head at the wrong time and made the winning bid. As the sale went along, the prices came down into my range and I bought my first two Beefalo heifers, Dolly and Loretta. Later, when I was back at the Opry for regular shows, I swear I could still hear mooing from backstage.

Who Let the Cows Out?

My daddy said, "Son, if you have equipment, it will break. If you have animals, they will get sick or die. And, if you have cows, they will get out, no matter how good the fences are." He wasn't being pessimistic, just realistic. So, I'm never surprised to get a phone call from a neighbor and hear, "I think your cows are out." I always go check on the situation, even if the description of the cows sounds like someone else's cows.

Grain farmers don't get calls in the middle of the night saying, "The fence is down and your corn has gotten out of the field." Every neighbor I've ever had has called me at one time or another with a cow report. Most aren't bothered about it. They just want to be helpful. But, when a subdivision was built on two sides of my farm and one of my cows got into someone's newly sodded lawn or flower bed, the caller wasn't happy and wanted action immediately. It was just as bad if my bull went through a fence into a neighbor's field of young heifers.

Of course, the most frequent occasion for the cows to get out is when I am out of town and my wife, Claudia, is the one to get the call. She's never happy to get roundup duties. Her first call like this came from the sheriff.

In the middle of the night, one of our black cows jumped the cattle guard on our driveway and was grazing on the roadside. A passing motorist saw the cow and, not knowing whose it might be, called the sheriff to prevent an accident. A black cow on a black night by a black road is a bad combination. The sheriff sent one of his deputies to our road to find the cow. He found her alright. He hit her with his car! He hit her hard enough to knock her down and do some damage to his car, but before he could get out of his car to investigate any injuries or damage, the cow got up and jumped back into the field. The sheriff was calling to report all of this to Claudia. He was a bit embarrassed to admit that his deputy hit the cow he was trying to find. I never heard from anyone about paying for any damages, so his being warned about a cow on the road must have kept me from being responsible. When I got home the next day and heard the report, I went to check on the cow, which was grazing in the pasture without a mark on her. Other than the deputy's car, the most damage was to the deputy's bruised ego. His buddies at the sheriff's department still tease him about running over the cow he was trying to find.

How a cow gets out isn't always understandable. If the fence is old or the wind has blown a tree across it, it's obvious why the cows are out. Or, if a gate has been left open, one usually knows who to blame and don't have to ask, "Who let the cows out?" I've had cows drink water from the river for ten years and they never tried to cross it. Then one day my neighbor calls and reports my cows are on his side of the river. Why did they suddenly decide to go for a swim? I've had cows get into my hay field and stand belly deep in lush alfalfa. Sometimes they damage more by just walking through the field than by what they eat. I've had to pay my neighbor for the corn they ate in his field. I never had to pay myself for what they ate when they got into the feed room, but was concerned they had eaten so much to founder themselves. The cows had to personally pay the two times they got into the hay field and knocked over the bee hives on the edge of the field.

The biggest mess I ever had was when I found them in my workshop. They clattered around as I chased them out. I had a clear picture of the expression, "Like a bull in a china shop." What they didn't knock over, step on and break, they splattered with manure. I was still finding manure on tools months later. It was everywhere! Why would a cow want to go into a tool room and relieve herself?

My most costly experience with the cows getting out was actually a getting-out-and-in episode. Our dog started barking in the middle of the night and wouldn't stop. I finally got up and let her into the house. I figured a nocturnal critter was walking across her territory and she was sending it a warning. The next morning, I discovered she had been barking at our entire herd of cows as they passed through our lawn. One was still grazing there when I went to investigate.

As I walked further, I began finding dead or dying cows everywhere I looked. One was dead in the driveway and another was dying beside my truck in the garage. As I kept walking to find the other cows, I found some grazing nearby, but still more dead ones. I had no idea why they were dead. I saw some dead ones near my neighbor's barn and went to investigate. My neighbor only raised hay; he didn't have any fences to keep his cattle in or my cattle out. When I looked into the barn, I found another dead cow and the deadly source. The cow had a grayish white material all around her mouth. It had come from a garbage can nearby that had some loose fertilizer leftover from a spring project. All fertilizers have a salty taste, so the cows found it quite appealing. The phosphate or potash in the fertilizer wasn't the culprit. It was the ammonium nitrate that killed them. When nitrates get into the blood stream, they create nitrogen gas, which is lethal in just a few hours. There is an antidote, but most veterinarians don't keep it on hand, because by the time the animal is discovered, it us usually too late. By the time I rounded up the herd's survivors and located the dead ones, I realized I had lost a fourth of the herd. To make matters worse, two of the cows that died had been national champions. The cows getting out this time had been costly. I later found they got out where a tree had fallen across a fence.

If someone's animal is going to get sick or die, they can be certain it will be their best one. When one of my national champion cows had her first calf, I had put her into a stall so I could keep a close eye on her. I wanted to make sure the delivery went without complications. All went well, and it was a handsome bull calf. I could already envision all of the money that would come in from breeding fees.

The next morning, when I went to check on them, the calf wouldn't get up. I examined him and found one of his front legs was broken between the knee and ankle. Evidently, the cow had stepped on his leg. The vet put a splint on the leg and for six weeks the calf, which we now called Tripod, cavorted around the fields with his mother. He grew like normal and I liked his looks more each day. This would be another grand champion! Unfortunately, when the cast was removed, the vet said the bone had not healed and would not heal. The only thing to do was to put him down. It's always the best one to get sick or die!

I usually understand why a cow has gotten out of or into something, but I'm no longer surprised when a new circumstance comes along. I've had cows eat polk berries from the pasture and die. My neighbor found a dead calf that got his head stuck in the fork of a tree. How did it get into that situation and why couldn't it get out? I had a calf born when the temperature was below zero, and his ears froze off. I've had cows whose tail broke and fell off.

One of my cows stole another cow's calf and raised it. I've seen cows drink from a puddle near the silo and go wobbling off because the puddle had alcohol in it that had drained from the fermenting silage in the silo. (Is there an Alcoholics Anonymous for cows?) I've sold an expensive heifer only to have her get so excited while she was being loaded into the trailer that she overheated and died. One of my cow's front feet hurt so much that to get relief she would get down on her knees to graze. I called her my praying cow.

I felt bad when I moved my cattle sixty miles from one farm to another and hauled a cow on one load, but left her calf behind. By the time I brought the calf on the next load, the cow had jumped the fence to go back the sixty miles and find her calf. I looked for days trying to find her, but finally gave up. About a year later, a neighbor two miles away found her bones and ear tag along the railroad track. Naturally it was one of my best cows.

To maximize the production of my champion cows, I flushed embryos out of them and implanted them into lower quality cows who would be the surrogate mother. Instead of a champion cow having ten calves in her lifetime, this allowed her to have hundreds. The breeding was done artificially. When a cow was ready to be bred, I would put her into a barn stall in the morning and call the inseminator, who would breed her at the proper time that afternoon. One afternoon he called me at my office and said there wasn't a cow in the stall. He wondered if I had forgotten to put her up. Since the stall walls were six feet tall, I wondered how she got out. I'm still wondering. Half cow, half deer?

My most unbelievable cow situation had a happy ending. It involved a cow being put into the barn for the inseminator, but this time I got a different call at my office. He said he had bred my cow but "she has an eye bolt in her eye." I asked, "What kind of bolt?" He said, "Eye bolt." Not understanding what sounded to me like a circle of conversation, I asked him to clarify, "I understand she has a bolt in her eye, but what kind of bolt is it?" He said slowly and distinctly, "She has an ....eye .....bolt ....in ....her....eye!"

After more conversation about the cow's situation, we hung up and I called the veterinarian. Just like the inseminator told me, I told the veterinarian that I had a cow with an eye bolt in her eye and needed him to come and get it out. Just like I had difficulty understanding the inseminator's report, the vet didn't understand mine. Taking my cues from the inseminator, I said slowly and distinctly, "She has an ....eye .....bolt....in....her....eye!" He got it!

When we got to the barn, there stood one of my champion cows (Did I say it's always one's best cows that get injured?) with an eye bolt dangling like an earring from her eye lid. This was no small bolt. The bolt was 3/8-inch-thick, and the circle of the eye bolt was two inches across. It was one of the eye bolts that was screwed into a post for a gate latch. As best we could determine, she was probably rubbing her head against the post when her eye lid got hooked in the eye bolt. Being caught, she must have butted the bolt with the side of her head with force enough to break the bolt off the post. The threads of the bolt were still in the post, and the eye bolt was dangling from her eye lid. I saw it and I still can't believe it! I would bet you couldn't break one of those bolts with a hammer.

The vet slipped the bolt from her eye lid and examined her eye and the bones around the socket, but the only damage he found was the pierced eye lid. After splashing some penicillin into the wound, he was on his way. I turned her out into the pasture, and she started grazing as if nothing had happened. I was left standing there trying to figure out how this could have happened- and to my best cow at that!

Even though I have seen a cow stand on three legs and scratch her chin with a rear hoof like a dog, it would create a panic if I called the vet and said, "I have a cow here with hoof in mouth."

The Cat

This cat was a "donation" from someone I knew, who was moving and didn't want to move their cat with them. That is translated, they didn't mind the cat having a move to my house, but they didn't want the cat moving with them. Translated, they didn't want this cat any longer. When I acquired this cat in 1984, it was a young cat, but I didn't really know how old it was. The owners didn't hang around long enough to pass along that information or its name either. I didn't really want a cat and had no attachment to it in the beginning; thus, it was referred to as The Cat.

My parents lived with me Monday through Friday operating my pencil business for a few months. Dad and The Cat took up with each other immediately. Dad would take a nap on the couch, and The Cat would lie on dad's chest and have a nap of its own. Like most cats he (by this time we discovered The Cat was a neutered male) liked being rubbed under the chin and along the jaw. He would stretch his neck out and close his eyes in ecstasy while being rubbed. If one rubbed his head with both hands, he would push his head through the hands so tightly that it pulled his skin back and gave his head a gargoyle appearance with accented eyes and disappearing ears. It would have been the same windblown look if he stuck his head out the open window of a speeding car. If The Cat wasn't rubbed just right, he would let one know. My brother and his family came for a visit in 1985 and his son, Drew, was playing with The Cat. Apparently, the rubbing was too rough for The Cat's pleasure, so he slapped Drew across the face and left a long scratch.

When Claudia and I married in 1988, she and Whitney had cat allergies, but since The Cat came with the deal, they were determined to adapt. Apparently, they adapted physically and socially because they stopped having an allergic reaction and quickly were accepted into the family. In 1989 Whitney went through several months of regular bed wetting. At the same time, The Cat started getting on Whitney's bed and urinating. We supposed The Cat figured, if Whitney was marking her territory, he would mark it as his. Claudia's sister, Kathy, diagnosed the situation. She asked us if we fed The Cat dry cat food, which we had been. She said it often causes cats to have urinary tract infections. We changed to canned cat food and never had any more urination problems out of The Cat. We could not get Whitney to eat the canned cat food!

Dry cat food or wet, canned or captured prey, one thing about The Cat's eating habits never changed. He always killed his food and then ate it. That would have been the expected routine if the meal was a captured field mouse. First, he subdued it, then played with it, then brought it around to show it off like a trophy. In the end, it always got killed with a violent shake of The Cat's head. He didn't have to capture, subdue, or play with the commercial cat food in his bowl, but he did have to kill the first bite with the same violent shake of the head just to make sure it was dead before he swallowed and took the second bite. I guess one can never be certain that the manufacturers didn't put beef, lamb, chicken or rice into the can or package while it was still alive.

About 1990, The Cat developed an open wound on top of his head and it became infected. The infection spread under his skin and down along his neck into a swollen pouch. We took him to our favorite vet, Dr Donnie Headrick in Franklin, who kept The Cat overnight. He called the next day to say The Cat was very sick. He wasn't sure what the illness was, but tests showed The Cat's white cell count was very high. Further tests revealed The Cat had an immunity deficiency. I told Donnie, "Are you telling me that my cat has AIDs?" "I guess I am," he said. "It's the first case in Williamson County!" In a few days The Cat was sent home in a cardboard carrying box with air holes. The box had printed on each side, "I feel much better now, thank you!" We felt much better too. The Cat was left with some scars on one side of his neck where the infection was drained.

Cats are known for having nine lives; and I don't doubt the claim. I think some humans may have even more than nine lives. At last count, I had survived twenty-six life threatening experiences. The least little thing seems to snuff out the life of dogs, cows, and horses. Frankly, the expression, "strong as a horse" must be limited to their pulling power and not their ability to fend off life threatening things like disease or accidents. Horses often get a mere stomach ache and die. A friend of mine says horses wake up every day looking for some way to die. Not a cat. They eat anything, go anywhere, do anything and still land on their feet.

I didn't chronicle The Cat's nine life episodes other than his battle with AIDs, but number nine was memorable and amazing. I don't know how old The Cat lived to be because he was already an adult when he came to live with us. He did live with us from 1984 until 1996. His final illness came suddenly. Whether through disease or injury, he lost use of his back legs. One afternoon I found him outside the back door using his front legs to drag the lifeless back half of his body along. The look on his face was not of pain or distress, but of purpose and determination. Since I had just arrived home, it was the normal time for The Cat to go inside and kill a can of lamb and rice for dinner. Before taking him inside, I gently examined him for signs of injury, but there were none. He didn't respond with pain to any of my probing. I guess if I had touched a tender spot, he would have slapped my face the same way he slapped Drew. He didn't hurt; he just couldn't use his back legs. After a brief drink of water and killing dinner, he stretched-out on a towel I put down by the freezer in the mud room.

The next morning revealed no change in his condition, but it was apparent he had been dragging himself around on the floor. After putting some food and water nearby, I left him in the mud room for the morning. Don't ask me to explain the how or why of what I am about to tell you. I can't. It defies reason.

When I came home for lunch, I was prepared for everything except what I found. I was prepared to find The Cat in worse condition or even dead. I was even hopeful to find some small change that I could label as improvement. What I found was no cat in the mud room. Could he have died and ascended into cat heaven? I didn't think so, but where was he? I looked under, beside and behind the storage chest and freezer, but there were not even signs he had tried to go to those places. Then I noticed the mud room door that opened to the outside. It was slightly ajar. He, no doubt, went out through this door. But how did the door get open? Had I not closed it completely? That was possible, but I thought I remembered closing it securely. Even if I had left it pulled closed, but not enough to latch, how could The Cat have enough two-legged strength to push it open? Could someone have come to the back door, opened it, and let the cat out? Possible, but I doubted that too. Even if someone had come and opened the door with no one at home, the cat in its condition wouldn't have scampered between the visitor's legs and escaped before the visitor knew what was happening. My best bet was on the door being ajar, regardless of the mechanics, and The Cat dragging himself out. But, why did he go out and where was he now? Had nature called and he headed for one of his natural outdoor litter box sites?

I looked around, inside and out, in all of his favorite hang outs. He wasn't snoozing behind the shrubbery or perched on the cellar roof outside the kitchen window. He wasn't in the stable, quietly waiting for an unsuspecting mouse to walk around the corner. He wasn't stretched out in the sun of the graveled driveway having a sun bath. He wasn't under the shade of the hackberry tree waiting for a pre-occupied bird to land nearby. He wasn't sitting on the front porch guarding the entrance and waiting for the first lucky family member to arrive home and get an affectionate meow welcome while winding his long supple body around their ankles.

Everywhere I searched I gave my usual meow call or greeting, hoping to listen and hear that familiar meow response. Neither of us knew what the other's meow meant at the time. It was just our verbal connection to be sent and interpreted as we wished. His may have been a "welcome home, where have you been, what took you so long." Mine could have been a "good to see you, I missed you, let's go inside." Or he could have said, "I'm hungry, feed me, now!" I could have responded with "OK, I hear you, give me a minute, I'll get something for you to kill and eat." After we stretched out on the couch and a had a rub under the chin, he might have said in a pitiful voice, "you can't imagine how awful it was" or "how was your day?" and we could have both responded with, "tell me all about it."

I widened my search and increased the volume of my meow calling of "can you hear me, where are you?" I went from the vicinity of the house to the garden, chicken house, pig house, adjoining hay fields, and the trees along Russell Creek. I looked and listened for any signs, thinking The Cat might have dragged himself away from the house, gotten tired, and was resting up for a return. The only sign I saw or heard was back at the house. The mud room door was ajar.

After lunch I went back to work. When I returned that afternoon, I repeated my search with the same empty results. The

Cat had vanished. Did a hawk, dog, coyote, or another one of a cat's predators taken advantage of his loss of mobility? Or had he done as I heard that some animals do, gone off to die? It's hard to lose an animal, but I much prefer to know who, when, where, and why, than to be left with a handful of unanswered questions. I would even rather find a dead body, than be left to sort through all the possible scenarios, some probably worse than the fact. I went to bed that night, woke up the next morning, and moved through the next work day sorting through the possibilities and gradually giving in to the fact that I may never know what happened to The Cat.

I don't remember why, but for some reason when I got home that afternoon, I went to the back of the house and looked through the opening in the foundation that was our access under the house. The old wooden door that covered the opening had long ago lost one of its hinges and frequently hung askew. I unlatched it, swung it open, poked my head inside and let my eyes adjust to the darkness. As the cool damp air hit my face, I gave out a meow that I hadn't used all day, and much to my surprise, I heard a meow response. I gave another meow to hone in on where the sound was coming from. I heard and then saw The Cat on a ledge about ten feet in front of me. Was that ever a welcome sight? In spite of the fact that he was still dragging himself along, he was alive and looked no worse. All this time I had been pacing the floor wondering where The Cat was and he was right under my feet hearing my footsteps. As amazed as I was that he escaped from the mudroom, I was more confounded by how and why he got from the mudroom to his position under the house. In the past he had regularly gone under the house to catch mice, get away from everybody, or cool off. The structure of the crawl space (no pun intended on The Cat's part) was not an easy maneuver for man or beast, healthy or unhealthy, but especially for a cat pulling itself along with just its front legs working. The crawl space of this 1840s house had doubled at one time as a place to store canned food, potatoes, and other things to protect them from freezing during the winter. It was also a cool storage spot in the summer. As soon as the door was opened, there was a ten-by-ten-foot pit about three feet deep that dropped from the level of the door. Somehow The Cat had gotten through the askew door, down the wall of the pit, and up the other side, all under the power of two legs.

I scrambled down into and across the pit to lift The Cat into my rescuing arms. We meowed back and forth multiple times, all with happy messages, I'm sure. I returned The Cat to the mudroom and attended his every need that night. I didn't notice any improvement or decline in his condition from the days before. The next morning, he was stretched out on his towel in the mudroom, having used up all of his nine lives.

He had loved to sit at the peak of our cellar roof above its entrance door. From this vantage point, he could view the fields of our farm and the approaching driveway. He could turn toward the house and see us at the sink through the kitchen window. Everything seemed in place with the world when we looked out the window and saw him there. So, it was fitting that we dug a grave at the cellar door threshold and buried him under his favorite perching spot.

Dog in a Baby Walker

Gone are the days of the dog pound and dog catchers. Now we have animal shelters and animal control specialists. But, stories of dogs with uncertain ancestry, coming from the pound and being the best ones, we ever had, still abound. My first dog as an adult came from the pound. but she definitely wasn't a Heinz 57 or Strooch (half stray, half pooch). It was clear from her markings that she was a purebred Irish Setter. I could tell she had been well cared for because her auburn coat shined like she had been brushed every day. The feather-like plumes on her tail and hind quarters looked like they had just been washed and blow dried. I could just picture her with a front leg raised and standing motionless on a point of a covey of quail at our farm. In one of my least creative moments, she was named Cinnamon.

When fall came, I didn't waste any time taking her on the first hunt. I had no clues about any training she may have had, which was probably just as well; because I didn't know anything about training a hunting dog. This would be a total test of her natural instincts. It didn't take long for the test results to come in.

We were walking through the pasture, no more than two hundred yards from the house, when she froze into that classic pose I had imagined. Her alert eyes and head were fixed on a clump of weeds about five feet in front of her. The left paw was raised and ready for the next step, but motionless. Only the feathers of her outstretched tail moved in the gentle fall breeze. I inched forward ahead of her with my shotgun raised and ready. Even though I know what to expect, I am always startled when a covey of quail is flushed and take flight. Compared to the soundless flutter of smaller birds, the fat round bodies of a quail taking flight reminds me of a lumbering 747 jet trying to get airborne.

I may have been startled, but Cinnamon charged into their roosting spot as soon as the first bird took flight. I shot into the middle of the covey and saw one of the birds drop to the ground about twenty yards away. Not wanting to lose sight of where it landed, I moved quickly to the spot. I wanted to see if Cinnamon knew about picking up the dead bird gently with her mouth and bringing it to me, instead of treating it like lunch. If not, I would do something to teach her about retrieving.

Finding the bird, I cradled the gun in my arm and turned to see what Cinnamon was doing. I looked at the spot where we had flushed the covey to see if she was inhaling the intense smell of the birds roosting spot, but she wasn't there. I turned looking in all directions thinking she had continued the hunt and was looking for the remaining birds. I didn't see her anywhere. I called for her a few times, but she didn't come. Not sure what to think about the situation, I headed back to the house. Cinnamon wasn't in the yard around the house, so I called for her again with no results. I looked into the tool room where she ate and slept, and there she was, not lying on her bed, but sitting in the back corner and trembling like the fall leaves in a breeze. She gave me a scared, sheepish look that told the whole story. "I can't be a hunting dog. I don't like the sound of guns." She wasn't gun shy. She was gun terrified! So much for my getting a great hunting dog from the pound. I would just have to enjoy all of her other great qualities.

And enjoy I did. She was always ready to go with me to the fields to check on the cows, whether I was walking or riding a horse. She would still come across the smell of quail and go into that classic point. I'm sure she was relieved for them to be flushed without that scary shotgun blast following. She was always ready to hop into the bed of the truck for a ride to the farmer's co-op. In the winter she would curl up by the fireplace and pose for a Norman Rockwell painting.

The next summer I learned that it wasn't necessarily the shotgun blast she was afraid of. I was in the living room during a pretty violent storm coming with a lot of rain, thunder, and lightning. Between claps of thunder and cracks of lightning, I heard a racket in the kitchen that sounded like wind was tearing part of the house off. What I found was Cinnamon dripping wet and standing in the middle of the floor with that sheepish, scared look I had seen before. "I don't like loud storms either," she said. She had jumped through a window screen to get into safety. Mostly out of anger at her jumping through the screen and to prevent her from having a chance to shake water all over the kitchen, I picked her up and tossed her back through the open window. Before I could close the window, Cinnamon jumped back through it and into the kitchen again. Not to be outdone, I tossed her out again. Even quicker this time, she jumped back through the window, but this time I caught her half way in. While I struggled to get her out and she struggled to get in, I saw the terror in her eyes and realized she wasn't about to go calmly to her bed in the tool room. I lifted her off the window sill and put her onto the kitchen floor into a growing puddle of water. Both of us were soaking wet.

While she stood there dripping more water onto the floor, I went to retrieve towels, one for her, one for me, and several for the kitchen floor cleanup. As I left the room, she got rid of most of her rain water with a shake that started at her head and rippled down her body to the end of the soaked plumes of her tail. How do dogs do that? Now the kitchen and I were even wetter. She still trembled as I dried her with a towel. The look on her face told me she was grateful. I gave in and let her stay inside. She gave me a lick on the face that made my anger evaporate and a wag of the tail that made me feel guilty for tossing a dog, afraid of loud noises, out the window, not once, but three times. As I removed the broken screen and closed the window, I decided it would be better in the future to just open the door and invite Cinnamon inside when a storm was coming.

One morning I went to the tool room to feed Cinnamon and she was still lying in her bed. She was usually up at daylight patrolling her territory and ready for breakfast by the time I arrived. When I poured the feed into her bowl and she didn't even raise her head, I knew she wasn't feeling good. I rubbed her on the head and encouraged her to get up, but she didn't offer to move. Maybe she had an encounter with a horse or cow and had been kicked. I checked her for an injury; but everything looked and felt normal. I lifted her to her feet, but when I turned loose, she just wilted and fell back down onto her bed. There was something seriously wrong, but I didn't have any clues and knew she needed to be seen by a vet.

Cinnamon had always been healthy, so the only time she saw Dr. Brogli was when he came to the farm for cattle work. Once a year he would give her a rabies shot while he was there. My friends, Coy and Jim, worked as receptionist and technician for Dr. Brogli. Coy was there to greet me when I carried Cinnamon into the waiting room. She knew Cinnamon from visiting us at the farm. Seeing I had an arm load, she led us straight back to an examination room where I put my limp-as-a-dishrag dog onto a metal exam table. After Dr. Brogli examined her and drew a blood sample, he said he didn't think she had an injury and thought the problem was a nervous system issue, but he didn't know what it was yet. He said I should leave her with him and they would so some X-rays and more tests to see what they revealed.

The next day Coy called to report that there was no change in Cinnamon's condition and the vet thought she probably had a viral infection of the nervous system. They had put her on an IV drip and steroids. I continued to make daily phone calls and frequent visits, but the report was always the same. All Cinnamon could do was lift her head. She was beginning to get bed sores where the constant lying on her sides created pressure points on hips and shoulders. The IV drip and daily steroids continued.

After two weeks of no progress, my wife and I had a difficult breakfast conversation and decided it was time to put Cinnamon to sleep. I drew the short straw and went to the vet's office to give them our decision. When I arrived, Coy said, "You have a walking dog!" I couldn't believe what I heard. Coy wouldn't joke about something like this, but I couldn't think of any way what she said could be true. "Well, she isn't exactly walking, but she did take a step this morning. We picked her up onto her feet, and she stood there with our support and then took a feeble step."

Even though this was a long way from having a healthy dog, I guess I could truly say it was a first step. I quickly forgot about the message I had come to deliver and didn't tell them my purpose for being there. They took me back to see Cinnamon and picked her up so I could see with my own eyes that what they were saying was true. I stroked Cinnamon on the head, and she acknowledged it with a slight wag of the tail.

The IV and steroid treatment continued with the addition of some creative physical therapy. Jim modified a baby walker, so a sling would support Cinnamon's weight under the mid-section and allow her feet to just touch the floor. In a few days Cinnamon went from just standing in this strange, but effective, apparatus to scooting all over the room at will. After two more weeks of physical therapy and steroids, they told me Cinnamon would be ready to go home the next day. She would still need lifting up, when she wanted to stand. After she squatted to pee, she would have to be lifted up again.

The next morning, I dreaded going to the vet's office almost as much as I did the morning, I went to tell them to put her to sleep. It wasn't my heart I expected to be hurting; it was my pocketbook. The cost of a month of treatment had never been mentioned, but I figured the bill would be gigantic. Jim led Cinnamon on a leash into the waiting room, while I pulled out my checkbook to settle the account with Coy. At least I would be able to see my walking dog and the benefits of the big check I was about to write. I was even prepared to ask for a plan of installment payments, if the bill was beyond our bank account balance. With a lump in my throat like I was about to say "put her down," I asked, "How much do I owe you?" Coy said, "Seventy dollars." Once again, I couldn't believe my ears.

"This can't be right. Just the syringes and IV supplies would be more than seventy dollars," I countered. They had been treating my horses and cows for a few years, but that amount of large animal business wasn't enough to offset this much of Cinnamon's bill. "Well, we've kind of grown attached to Cinnamon around here."

I paid my bill and asked to see Dr. Brogli. I wanted to thank him personally for his good care and generosity. They took me into a surgery room, where he was in the middle of operating on a big Siamese cat. Taking care of people's animals was what he was all about and this experience would make me forever grateful and indebted to him and the other vets I ever worked with.

Cinnamon's bed was moved from the tool shed into our kitchen, where I continued to give her daily medication and treat her bed sores. For a couple more weeks I still had to lift her from her bed and lift her again, when she squatted to pee. Gradually all of her ailments improved and she was back to running in the fields and hopping into the truck for a ride to the co-op.

We soon moved about twenty miles away to a farm near Franklin and found a new vet to take care of our animals. I told Dr. Headrick about Cinnamon's history with Dr. Brogli. He said nothing about the stay surprised him except the great recovery.

I had guessed Cinnamon was about three years old when I got her from the pound, so she must have been about ten by this time, which was a pretty normal life given all she had been through. On one of her regular check-up visits with Dr. Headrick, he noticed a lump on her neck and said it was probably a cancerous tumor that would just have to run its course. The tumor grew slowly and didn't seem to bother Cinnamon, but after about a year, she began to decline in condition and we decided one weekend that once again it would be best to put her to sleep.

I drew the short straw again and took Cinnamon to the vet on Monday morning. (Beware if I volunteer to take you to the doctor on a Monday morning!) As fate would have it, Cinnamon was more frisky than normal that morning and seemed to feel better than she had in weeks. I was already second guessing our decision.

I didn't have an appointment and when I led Cinnamon into the waiting room, it was a packed house. Cinnamon was always well behaved around other animals, even cats, so she sat quietly beside me and leaned against my legs. I didn't need this extra time to think about our first and last hunting trip, broken screens, rides to the co-op, baby walkers, or what I was about to do. It didn't help any that Cinnamon gave me an occasional lick on the hand. As she looked around the room at the other animals, a lick was her way of saying "I belong to you. I like you."

Finally, all of the other patients cleared away and Dr. Headrick came into the waiting room to greet us. Normally, he would have the receptionist take us back to an exam room and see us there, but he knew why I had come. Being as good with pet owners as he was with their animals, he noticed I couldn't talk, even in response to his friendly greeting and asked, "Do you want to leave Cinnamon with me?" I opened my mouth to say yes, but the word wouldn't come out. I just nodded my head, handed him the leash, and walked out.

Black Angel

I'm not very creative when it comes to naming pets. If it's a brown dog, name him Brownie. If he has spots, name him Spot. When Lane dies, name the next dog Lane. If the cat goes a long time without a name, just stick with The Cat. In an attempt to be sophisticated nine-year-olds, my neighbor and I named a stray dog William. Some names just weren't meant for pets.

Reverse thinking must have been on my mind when we named our Great Dane pup. She was sweet and cuddly like most pups, but Angel wasn't the first thing that pops into one's mind when seeing a black dog. What were we thinking? Not much, obviously. Neither did we think things through on the costs of purchase, failed ear clippings, spaying, and annual vet bills for a breed of dog whose average life expectancy is seven years. As much as we enjoyed Angel over the years, I don't think I'll buy another Great Dane with those odds. Even if I do, I'll pass on the ear clipping.

Angel may have started as a pup small enough to sleep on my lap, but I swear, I could almost feel her growing as she lay there. She wasn't a lap dog for long. I built a wooden box and put it in the barn feed room for her to sleep in. She soon grew so big she had to curl up to sleep in it. She would soon become nearly three feet tall at the shoulders, weigh 150 pounds and big enough for children to ride. In no time I was buying dog food at the farmer's co-op in five hundred pound lots to get the best price. Her wagging tail felt like a whip if she hit someone with it, and she could rear up on her hind legs to eat apples from the tree. Leaning against anyone she was quite a force, but let her run between someone's legs and they would be upended. I had to quit wrestling with her on the floor after she punched me with a right cross to the face, that made me think I had been hit by Muhammed Ali. As soon as I cowered in pain, her sweet side took over and the rough play stopped, as she licked my swollen, soon to be black eye.

Angel's size both amazed and terrified people. The Cofers visited us with their three-year-old son, Adam. Angel outweighed Adam four to one and came up to his chest. They looked at each other eye to eye. After they had played in the living room for a while, I opened the door to let Angel go outside. Adam stuck his arm out to pat Angel good bye as she dashed past. The result was Adam being spun around and nearly losing his balance. With an astonished look on his face, Adam said the obvious. "Big dog!" From then on, he would ask, "How's Big Dog?"

Our daughter, Whitney, was seven years old when she first met Angel. She was familiar with dogs, but her border collies, Serve and Volley, were not in the same class as Big Dog. It didn't help that their first meeting came as she was stranded half way between the car and the back door of the house. This big black creature came around the corner of the house and headed straight for her. Her screams only excited Angel more. Whitney tried to get to her mom for safety, but there was no way around the prancing Angel. Our words of "he won't hurt you" were of no reassurance. Whitney wasn't the last to be terrified of Angel.

The FedEx driver wouldn't get out of his truck, when he saw Angel in the driveway. He just blew his horn until we came out of the house. With his window rolled down just enough to yell through the crack he called, "I'm not getting out of this truck. If you want your package, you will have to come and get it." There was no convincing him that this big black dog was an angel.

My cousin, Mary, parked her compact car in our driveway, but before she could get out, Angel had her front paws on top of the car and was looking down at Mary through the open window.

Another spring afternoon I was working in the yard when I looked up to see three boys walking across the field from the elementary school next door. Ball gloves in hand, they said Angel had come over to the playground and taken their softball. Angel was standing nearby, but no ball was in sight. I told the boys I didn't know anything about their ball. They insisted she took it. Angel gave us all an "I'm innocent" look before opening her cavernous mouth to let their now slobbery ball drop to the ground. The boys grabbed the ball and ran back to their game. I later learned Angel was a frequent visitor to the school and would saunter through an open classroom door.

Angel wasn't a barker. When a stranger came around, she would just give them a stoic look and make them guess what she was thinking. No growl to say I'm mean. No tail wagging to say let's be friends. If I was suspicious about the stranger, I would use the uncertainty about Angel to my advantage. They might ask,

"Will that dog bother me?" I would reach down to hold Angel by the collar and reply, "Only if I want her to."

I didn't have time to use my line on two teenage boys I spotted dumping brush into a sink hole at the front of our property. They were unaware as Angel and I walked the hundred yards from the house to their pickup truck. As usual, Angel paced like a horse ahead of me. One boy was in the bed of the truck tossing limbs down to the other boy on the ground. Angel was beside the boy on the ground before either of them knew anyone was around. He wasn't just startled, he was terrified. He screamed at me, "Don't let him get me! Don't let him get me! Please help me mister!" He jumped into the truck bed for safety. It wasn't safe enough because Angel made an easy jump into the truck to join the boys. To their relief, I called off the man-eating dog before she could lick them to death. They wouldn't be coming back to dump brush on our property.

Even if we told visitors in advance about the big black harmless dog, they weren't convinced after seeing her for the first time. My neighbor, Rich, kept a horse on our farm and would bring friends over to ride. He told one friend to just hang out at the barn, while he went into the pasture to catch the horses. Angel wasn't around for an introduction, so Rich told his friend to not be alarmed if a big black dog came around. "She's big, but harmless," Rich said. When Rich got back to the barn with the horses, his friend wasn't around. He looked in the barn, shop, and tack room, but still couldn't find him, so he went to the house where the guy's car was parked. The friend was in his car with all the windows rolled up and the doors locked.

"What are you doing in here?" Rich asked. The friend replied, "I thought you said that dog was big, but harmless. While you were gone, it came around the corner of the barn dragging a dead calf by the leg!" Rich just laughed, when he realized Angel had been to the farm next door and found a still born calf and dragged it home. As far as Rich's friend knew, Angel had cut a calf out of the herd like a lion, killed it, and brought it home for lunch.

When we bought the three-hundred acres and a country inn, Angel had even more space to roam and more visitors to amaze. She liked to walk around with guests and would sometimes follow them into their rooms. Jennifer, a desk clerk, would get a call from a guest saying Angel didn't want to leave their room. Jennifer would get a leftover breakfast biscuit from the kitchen and lure Angel outside.

The guests didn't always report Angel's room visits. One morning while I was helping serve breakfast, I heard bits and pieces of a conversation among a group of businessmen. After a while I caught enough of the story to learn they were talking about Angel. It seems Angel had followed a man back to the Jack Daniel's room where he was staying. He said she came into the room, circled twice, and laid down. He tried to call her out, but she didn't budge. Afraid to grab hold of her and pull her out, he just turned out the light and went to bed. As he told this story there was plenty of laughter at his expense. One friend said, "I couldn't have done it. I couldn't have slept in there all night. The smell would have been more than I could take. Yeah, I don't understand how that dog was able to stay in there with Fred all night!"

We were accustomed to Angel going missing for a few hours only to show up from a walk around the farm with guests. When she was gone overnight, we went looking for her. Once, we looked for several days and couldn't find her. We visited all of the neighbors and scoured the farm looking for places she might have gotten hurt. Had she wandered away? Did someone steal her? We had exhausted all our possibilities and gave up looking. A week after she disappeared, I was at the barn and walked by the scale house. I had already searched the barn area several times for her, so I was no longer looking to find her. I had walked past the scale house several times in my search, but this time I just happened to look down into the scale pit and was shocked to see Angel at the bottom of the six-foot pit. She was alive! From the blackness of the pit, about all I could see was the whites of her eyes, which seemed to say, "I heard you all of those times you came walking by." My mind was flooded with thoughts and emotions as I lowered myself into the pit to rescue her. How many times had I walked past this spot and not looked down to see her? There was no food or water down there. How had she survived all these days? Fortunately, it was like a cave and the temperature was a moderate fifty-eight degrees. I was very happy to see this big black dog, even if she was thin, hungry and thirsty.

In a few weeks Angel was back to her old form, but as she passed the seven-year life expectancy milestone and on to eight, nine, and ten years old, her health declined noticeably. The first symptom was no longer being able to jump into the back of the truck for a ride to the farmer's co-op. She would put her front paws onto the tailgate and wait for me to lift her hindquarters onto the truck. She loved riding in the back of the truck with her head stuck around the edge of the cab. The wind blew her poor performing, floppy, clipped ears. She always attracted attention from people in neighboring cars at the stop lights. With the truck backed up to the co-op loading dock, she could walk right off the truck and into the warehouse to inspect their dog food supply.

Soon she wasn't able to eat enough dry dog food to maintain her weight and we started feeding her with canned food. I bought ten cases at one time. She still couldn't eat enough to maintain her weight and became embarrassingly thin. We were sure the inn guests thought we weren't caring for her properly, so we started keeping her at our house all the time. Even then, she would make her way the half mile down the hill to the inn where all of the people were. To stop this wandering, I converted two spacious horse stalls into what we called Angel's nursing home. When we were going to be around the house, we let her out onto the lawn and into the house in cold weather to sleep with The Cat.

One evening I was going to put her into the nursing home for the night and I couldn't find her. I looked all around the house and lawn, but there were no signs. I went down to the inn to see if she had heard guests and gone down to be with them. I didn't find her and none of the guests had invited her into their room or even seen her. I went to bed thinking she would show up the next morning, but she didn't. I expanded my search and of course the pit at the scale house was one of the first places to check. After several days of fruitless searching, we gave up. I had heard of animals often going off to die and thought maybe that's what Angel had done too. As the months of fall and winter passed, out thoughts of ever finding Angel alive had stopped, but we still wanted to know what happened to her instead of filling the unknowns of our minds with what-ifs. Six months after she disappeared, I walked the half mile down the hill from our house to the inn. As I walked along a fence row about two hundred yards from the inn, I came across a skeleton with its back against the fence. Angel had been on her way to the inn to be with guests, got tired, lain down, and died. It was sad to know she ended her thirteen years like this, but at least we had an end to our wondering what had happened. This angel spent her last day on her way to be in a place she had spent many happy days and nights.

Right Lane, Left Lane, Seldom Wrong Lane

Lane was one of my favorite dogs, probably because she was my constant companion, especially if I was going somewhere in the truck. You can relax; this isn't going to be a story filled with Lane's near misses or tragic death; even though I had a lump in my throat that wouldn't go down for quite a while, when she died unexpectedly. Like most pet owners, who lose a special animal, it wasn't unusual that I had unexpected tears as I walked across the Wal-Mart parking lot and realized Lane wouldn't be waiting in the truck for me.

Lane was a red Golden Retriever that we bought from a family named Lane. With that kind of creativity, it's a wonder our daughter wasn't named Baptist after the hospital where she was born! Instead she was named Whitney after a friend's cat. Lane was supposed to be Whitney's dog or at least the family dog, but since she spent all day with me, I sort of commandeered her from the rest of the family.

Regardless of the allegiance to any one of us, she would be jealous if someone was giving attention to someone else instead of her. When one of the family arrived home and was greeted by the others, Lane was there to greet too. If we exchanged a hug or a kiss, Lane would move between us to break it up. If one of us was rubbing on Lane, she would turn her head to the side and look at the others in the room, as if to say, "See, I'm getting all the attention. He's all mine."

Lane was the prettiest when we could watch her from our breakfast room window; as she patrolled the green grass of the hay field. Her shiny red coat was a perfect contrast of color with the green grass. The morning sun would catch the golden feathers of her tail and hindquarters and make them sparkle. When she finished her patrol, she would return to sit on the deck, looking over the landscape, like a queen admiring her dominion.

When she looked at someone with her mouth slightly open and tongue out a bit, they would have thought she was smiling at them and posing for a Luck's Bean commercial. Guests at our inn were always asking her for the secret recipe.

I guess physically she only had two flaws. One was her teeth. She could have benefited from some time at the orthodontist. If she had eight front teeth, they pointed in nine directions. Her other nagging affliction was being allergic to grass. That's unfortunate for a farm dog. She scratched a lot, which made people think she had fleas and negligent owners. Her favorite allergy relief was to lie on the rocks of our driveway and roll around on her back and on both sides, moaning with pleasure. The result gave us pleasure because she created "rock angels" where she had been writhing and rolling.

Lane let us know in just a few months that her retriever breeding wasn't just a hunting dog heritage, but that she had true bird instincts. The first few months of her puppy life were spent in our house or out in the yard. She had been good about staying put when we went away for a few hours and left her in the yard. Even though our closest neighbors were a quarter mile away, I checked with them occasionally to see if Lane had been coming to their house and the report was always that she hadn't. I was glad to hear that because I didn't want her wandering the neighborhood.

We had gone to the inn one morning to do some outdoor chores and left Lane in the yard at home. We had been working for about a half hour, when I looked up and saw Lane galloping across the pasture toward us. She had never come from the house to the inn on her own. It was at least a half mile down the hill, through the creek, and across the pasture for her to get to us. But she must have heard our voices and wanted to come to be with us. In later years I trained her to come to the ringing of a bell. If I had left her at the inn and wanted her to come to the house, I would ring the bell and she would either run the one mile along the driveway or take the short cut across the creek and up the hill. Unfortunately, when she chose the creek route, I was greeted by a wet dog and had to towel dry her before she could come inside.

After her first run down to the inn, we greeted and praised her for finding us and then went back to our chores. Lane went off to explore the inn property. It wasn't but five minutes before she made a discovery she really liked. We heard a commotion from the inn back lawn and then a squawk. Lane had found Sneaker, our rooster that roamed free around the inn. The squawk was the end of Sneaker as Lane did what bird dogs do to live birds and I don't mean retrieve.

Lane would stop anything she was doing for a chance to ride in the truck. I could yell "Truck!" and she would come running from anywhere she was and jump into the bed of the truck. If the weather was bad and I wanted her to ride up front with me, I would say "Cab!," and she would race to the truck and wait by the door for me to open it.

I was actually surprised she became so attached to the truck given one of her first experiences with riding in the back. I had loaded a bunch of trash onto the truck bed and was going to drop it off at the dumpster on the way to taking Whitney to school. Lane wanted to go too, but with three people in the cab, I chose to put Lane on top of the trash. I would drive slowly, so she wouldn't get thrown off. We were less than a mile from home when the load shifted and Lane decided to abandon ship. She jumped out onto the roadside. We weren't moving but about twenty miles an hour but Lane went rolling in the grass when she landed. She wasn't injured, but it literally knocked the shit out of her. Whitney picked her up and put her into the cab with us for a smelly ride to school. It took quite a while for Whitney to get herself cleaned up after she got to school, but she still got some strange looks during the day.

Since it looked like Lane was going to be a truck companion wherever I went, I wanted to teach her to stay in the bed of the truck while I was inside a business. When she was six months old, I was working to train her to "Stay!" and she did pretty well with it. After some practice runs on the farm, I decided she was ready for the test on a trip to town. I got out of the truck at Lowe's and gave her the "Stay!" command reinforced with my hand up in the stop position. She sat down in the truck bed with her head hanging over the side, just as was expected. She gave me that longing look, but didn't offer to move from her spot. As I walked to the store, I looked back a couple of times and she hadn't moved. I had been in the store about ten minutes, when I heard someone laugh and I looked up to see Lane galloping down the long aisle toward me with that happy "I found you!" look on her face. I didn't know if the automatic doors opened for a dog or if she waited for her opportunity when someone was entering. I was more surprised that she found me in that big store, than I was that she got out of the truck. No doubt she just used her sense of smell to track me down.

Lane was equally happy riding in the back of the truck or up front with me. "Let's go!" may have been the favorite words for her to hear. It always made her hop up and down or race ahead. Where we were going made little difference to her. Saying "Car!" excited her too, as she raced to the car and waited for her chauffer to open the door.

I say the destination made no difference to her, but she definitely had some preferences and would let me know. If we were leaving the inn, when we crossed the wooden bridge on the driveway and came to the fork in the driveway, she would lean back against the seat as if to brace herself for a sudden turn. If I turned left to go up the hill to our house, she would just relax and scan the roadside for any critters that might run in front of us. When that happened, she would lean out the window as if to say, "Let me at him. I could catch you if I weren't in this truck!" If I turned right at the fork to head toward the highway, she gave an excited bark which said, "Oh boy, this is going to be a long ride!"

When we drove two miles down the highway and approached the entrance to our restaurant, she would again lean back and brace herself for a turn. If we kept going straight, she just relaxed as we passed the entrance, but if I turned into the entrance, she would bark in anticipation of going to the restaurant which was in an old mill on the river. It meant she could go for a swim while I was taking care of business inside or maybe she could find some young boys to play with. The fishermen weren't as happy to see her and have their fishing ruined.

Riding in the passenger seat of the cab, she kept her eyes focused on the road ahead. If she was in the back in the bed of the truck, she would put her paws on the top edge of the bed and lean out, so she could stick her head around the cab and see what was coming from ahead. If we met a car, she just kept staring straight ahead. If we met a big truck, she was obviously upset and would bark until it was out of sight, then resume her surveillance of the road ahead. I never figured out what it was about trucks that upset her, but she obviously thought they were a menace to the highway. To make it more confusing, she would occasionally have the same reaction to a particular car. Size, color, speed, sound? I never discovered a pattern, but she knew.

When cars or trucks arrived at the inn, she never bothered to chase or bark at them, except for one. The UPS truck was always greeted with a chase and a ferocious, "I'm going to eat you alive!" growl and bark. When the driver parked and got out of the truck, Lane was all sweetness and light. The driver was one of her favorite buddies, even if he wasn't delivering a box of treats for her. But, when he got back into the truck and drove away, she again became a momma grizzly protecting her cubs.

Mondays were one of Lane's favorite days for a truck ride. That was bank deposit day and she could always count on a dog biscuit from the drive through teller. Lane sat patiently in the passenger seat while I was giving the deposit to the teller, but when it was time for my receipt to come back from the slide out drawer or down the tube, lane would stand at my shoulder drooling in anticipation. I ended up with slobber on my hand and tried my best not to get any on the return canister, but I'm sure I wasn't always totally successful. The next patron probably wondered what they had gotten their hand onto. If the bank happened to be out of dog biscuits, it created a crisis that Lane didn't understand; so, I started carrying a few emergency dog biscuits in the truck. It's a good thing I did because the bank's corporate office issued a no dog biscuit policy and I had to provide my own to support Lane's habit. A corporate lawyer probably dreamed up some liability exposure for distributing suckers to kids or biscuits to dogs. Or a bean counter thought it was a useless expense that could be cut. Lane and I threatened to change banks, but the policy went away after one of their countless mergers and the dog biscuits returned. Lane started thinking every drive through window should pass out treats, but the folks at the electric company and McDonald's didn't understand the practice.

I did provide McDonald's with an experience for their staff to talk about. After placing my order, I asked the clerk if they had any dog biscuits, even though I knew the answer would be no. After a polite turn down from the clerk, I turned to Lane and told her they didn't have any dog biscuits. Lane didn't know the words dog biscuit (I'll explain that later.), so she gave me a blank unconcerned look. With the clerk listening, I asked Lane, "Do you want a hamburger instead", and then softly "or a cookie?" She knew the word cookie, (I'll explain that later too.) and hearing it always produced a bark. So, I turned back to the clerk and told her that Lane wanted a hamburger, but hold the pickles. Of course, the clerk was trained to follow-up with, "Do you want any fries with that?" even if it was a dog order. So, I turned to Lane and asked her if she wanted any fries. None of those words meant anything to Lane, so she just gave me a bored stare. "No fries, thank you," I told the clerk. Since we were in the drive through, I didn't get a "Is this for here or to go?" Lane made her junior hamburger disappear in one gulp before we could get out of the parking lot. Maybe McDonalds should put dog treats on the menu.

I promised to explain the "dog biscuit" and "cookies" words in Lane's vocabulary, so here it is. Somewhere along the way I learned that the average dog recognizes about a hundred and twenty words. Since we knew Lane had an above average intelligence, she must have known more than that. I would tell the inn guests that there was a problem with owning a Golden Retriever. "A man shouldn't own a dog that's smarter than he is," I would say. We counted thirty words that Lane knew, so we figured there were at least ninety more that she hadn't taught us yet. Cookie was a word she taught us.

One evening Claudia was in the kitchen cleaning up after supper, while I watched TV in the breakfast nook and Lane sprawled on the floor napping. "Do you want a cookie?" she called to me. Before I could answer, Lane sprang to life and dashed off to the kitchen to sit in front of Claudia. Lane gave her an expectant look. "Do you want a cookie, Lane?", she asked. Lane licked he lips and thumped her tail on the floor in a response that could only be interpreted as a yes. We weren't accustomed to sharing our food as treats for Lane and had no memory of ever giving her even one leftover bite of a cookie. Maybe she had overheard us saying cookie, watched us enjoying eating them, and assumed she would welcome the chance to have one too. Whatever the reason, she was ready for a cookie. Knowing this wouldn't be the last time she would want to get in on the cookie eating, we decided it was best to give her a dog biscuit and let her associate that with the word cookie. Lane had her own jar of dog biscuits sitting next to the cookie bin, so she could have her own cookies. If we didn't think it was the right time or place for Lane to have a treat, we had to just say "C" instead of cookie. More than once I saw her cock her head and look inquisitively at someone who had used the word cookie in a conversation. We resolved that when we got the next dog, we would start early on teaching it the one hundred twenty words we wanted it to know.

Physical signals were just as easily understood as words. As I moved around the house, she would follow me from room to room, watching for a signal. If I went into the study to write, she would follow me and lie on the floor by my chair. If I went into the living room to watch television, she may lay by the fireplace for a nap or across the doorway to make sure I didn't go through it without her knowing. If I was just going from room to room, picking up things I would need during the day, she followed me every step. Telling her I wasn't about ready to go and she should just lie down and relax had no effect; she kept following. She wasn't going to take a chance on me slipping out the back door and leaving her behind.

Putting on my jacket and zipping it up was a signal to Lane that we were about to go somewhere. My getting dressed in a suit was a signal that she probably wouldn't get to go, and she would go lie on her bed with a sad look. If she was outside and I went to the door to invite her into the house, she would refuse to come in because she had learned it was a definite sign that she would be left in the house while we were gone. She wanted to take chances on getting to go along by begging to get into the car when we were about to leave. Sometimes she won.

Lane also learned on her own by observation what a door knob was for. It was another one of those things she taught us. I don't mean she taught us the purpose of a door knob. We knew that! What we didn't know was that she knew. In this case it wasn't the word knob; it was what the knob did. Like other times when she taught us things, we learned quite by accident. I was leaving the house and, of course, Lane was between me and the door. Before I could open the door, Claudia called from the other room to tell me something. Thinking it would be a quick exchange and I would be on my way, I held onto the door knob. Lane was prancing with expectancy at the opening of the door. When I realized it was going to be more than a quick exchange, I took my hand off the knob and Lane sat down and relaxed. Sensing the end of the conversation, I grabbed the knob again and Lane resumed her prancing. Seeing this reaction, I thought I would test to see if the knob really was the cue. I took my hand off the knob and she relaxed again. This time I touched the door, but not the knob. Lane just sat there. I touched the knob and she was back to prancing. Lane couldn't use the knob to open the door herself, like some dogs I have heard about, but she knew nothing would happen to open the door until someone touched the magic knob.

Lane had her own club chair in my office where she dozed as I worked. At the end of the day, when I turned off my computer, it made a winding down whining sound that was Lane's signal that the work day was over and we were about to leave. With no other cue, she hopped out of her chair and stretched while the computer closed down and I finished my end of the day tasks. I've often wondered if a dog is smart enough to know it should stretch every time it gets up, why can't humans be as smart and stretch too?

Handouts got to be a regular thing for Lane at the inn. The desk clerks went so far as to buy a jar labeled "Lane's Treats – Stay Out!" The kitchen handouts were so regular that Lane gained a lot of weight and I put a ban on the kitchen handouts. They still slipped a leftover biscuit or piece of ham when I wasn't looking. I don't know how they managed to do this without me catching them. They said they could always tell where I was, because Lane would be nearby.

Lane didn't have a work free life at the inn. There was more to do than lie in the office chair and eat an occasional handout. She took on the job of greeting the guests at the front door. She was such a hit with the guests, that I gave her the title of Director of Public Relations. She was worth every penny of her compensation, which amounted to room and board, plus a zero-co-pay health care and drug plan. I gave her a column in the newsletter entitled "Lane's View." which gave guests a report of what was happening on the property from her perspective. The guests were more interested in what appeared in Lane's column than what I had in mine. One column had the heading "I'm Lucky" beside her standard picture. She went on to "write" all the reasons she was lucky to live at our farm and inn. Reading the column, the new guests thought her name was Lucky and would call her that when they arrived. I didn't try to correct the confusion my editorial impression had left. When she announced in her column that her birthday was coming up three days before Valentine's Day, the boxes of dog treats and birthday cards started filling the mailbox.

An announcement of my birthday didn't produce the same response.

When an email was sent to my advisory board about a financial report or an upcoming board meeting, it always came from Lane. With this added responsibility, I promoted her to Director of Communications. The board members told me point blank, that they hardly ever read what I sent, but would read every word from Lane. Most U.S. presidents seem to have a dog. Maybe they could take a lesson from Lane and have their dog do all of the communicating with the citizens. That would give new meaning to a yellow dog Democrat!

Lane's favorite guests weren't the ones who gave treats to her, but the ones who brought their dogs. Pets were welcome at our inn, as long as they were at least a year old. Even God can't control a pup! Lane would run, romp, wrestle, and play with the visiting dogs all weekend and by Sunday night she was exhausted and in her bed by six o'clock. She slept in a corner of our bedroom on a foam pad covered with a quilt. The only time she was on our bed, and even then it was uninvited, was in her later years when she became terrified of thunderstorms. She was our weather alert system. She would come to one of us at the side of the bed panting and whining. She panted like she had been running for miles. If that person didn't respond, she would go to the other side of the bed. Back and forth she would go until we couldn't take it any longer. To try and calm her down, we tried putting her into the bathroom or a closet, but none of those worked. She just barked until we let her out. Finally, she would jump into the middle of our bed and stand there panting and looking frantically about the room, as if something was going to come through the door or windows and do us harm. We got her to lie down between us, but that really didn't help. She kept up the panting until the storm subsided and she could return to her bed.

"Early to bed, early to rise" was Lane's motto. If we weren't already awake by 5:30 a.m., Lane would come to our bed and poke an arm, neck, or face with her cold, wet nose. It was her message that it was time to get up and let her go outside. If this rude awakening didn't get the desired response from the first one, she would go poke on the other side of the bed. If that didn't work, she was persistent and barked as if to say, "If you don't get up and let me go outside, I'll just pee here on the carpet." I've read that dogs can control their bladders for a lot longer than humans, but we were afraid to put it to the test. Besides, with all of the barking, there would be no more sleeping. So, one of us would get up and let her out. There was no need for us to stumble back to bed, for in a few minutes here would be a bark at the door saying, "Open the door, I'm ready to come back in for breakfast." Her food bowl and water bowl were in the corner of the breakfast nook and, until she started putting on weight in her later years, there was always food in her bowl for her to eat on her own schedule. When we happened to let the bowl get empty, she would stand beside it while we refilled it and then just look up at us until we walked away. She wanted to eat in privacy. "Go away, I don't stand around your table watching you eat," her look would say.

We never fed scraps to Lane from our table, but she got plenty of other supplements to her diet that we had no control over. I've always been puzzled by people who are passionate about not giving a chicken bone to a dog. I'm sure somewhere along the line, some dogs got choked when one of the splintered bones got stuck in the throat.. When that happened, a new rule was established for all dogs. I was helpless to completely enforce the rule with Lane, because she regularly went into the field and caught a bird. Her eyes were always alert to anything feathered; after all, she was a bird dog. The retriever part of her name didn't refer to her bringing back a tennis ball, that had been tossed into the yard. She caught birds, rabbits, moles, and even an occasional squirrel. Once I saw her tangle with a fierce groundhog and win. We didn't have to buy toys for Lane. She caught her own.

When Lane caught an animal in the field, she retrieved it to the yard, sometimes still alive. She wanted to show off the booty from her hunting skills. "Look what I caught." She might lie down with her prey in front of her and one paw on top of it to keep it from getting away. She would ease her paw off of it to let it escape and then grab it again before it got away, once again proving the point that the chase was more important than the catch. Like a child, she liked to play with her food. Eating it wasn't the main purpose. It soon became a toy. Days after the capture we would see her in the yard with a bird or a mole, still playing with it. No part of it had been eaten. She would grab it with her mouth and toss it into the air, hoping it would take flight again or hit the ground and try to run away so the chase could be repeated.

Lane had two periods of time in her life when there wasn't enough time or energy for being frisky and playful. The year she was two she had two litters of pups. Both were planned pregnancies – sort of. Claudia wanted to raise puppies. That's why she chose a female pup when she bought Lane. She found a male Golden Retriever, named Rusty, in the nearby town. But when the pups came, the litter was red, black, white, and spotted. No doubt, Rusty wasn't the only father. It would be possible, but strange for a woman to have fraternal twins with different fathers, but I've never heard of it. It happens a lot with dogs. So much for mating for life in the canine world.

We were careful with the next breeding. In the first days of Lane's heat, we put her in the house mud room. It was smart because dogs started assembling at the back door at all hours. They jumped onto the door with such force that I thought they would break the door down. We had never seen any of these dogs before and had no idea where they lived. Need evidence of a dog's great sense of smell? This was one.

In a few days we invited Rusty over for a long weekend and locked him and Lane in a stall in the barn. The next morning there were three dogs in the stall. Lane, Rusty, and a male mutt I had never seen before all greeted me when I opened the door. The mutt had knocked boards off of a stall window that was four feet off the ground and climbed through. Now that was determination! I didn't know anything about this mutt's parentage, ownership, or residence, but I had evidence of his physical strength. I had different hopes for this breeding, but feared we were in for another mixed litter. My fears were well founded.

Lane was an attentive mother to her black and red pups. She raised both litters without complication. When the spring litter was old enough, they played outside under the smoke house. Looking for objects to practice their chewing skills, they attacked a newly planted snowball bush at the corner of the smokehouse and chewed it down to a single six-inch stick. The bush miraculously survived and is now six feet tall. Lane grew weary of the nursing duties before the pups did. I'm sure the developing teeth didn't help any. Instead of lying on her side for them to nurse, Lane walked by the pups and they would rear onto their hind legs, reach up to the teats and nurse. We gave Lane the nickname, Stand-N-Snack.

It was always a welcome sight to arrive home, pull into the driveway, and see Lane waiting expectantly on the front porch. It was a low porch that was only one step up from the sidewalk. Lane would often be sitting with her rear end on the porch and her front feet on the sidewalk as if her rear end needed a rest, but her front legs were ready to spring into action. Our getting out of the car was one of those reasons for springing into action. It often prompted her to run laps at full speed around the yard, as if the only way to control her excitement was to burn off her pent-up energy. It was nice to arrive home and know someone was that excited to see me. However, I suspect that if Claudia arrived home and I started running laps around the yard at full speed, she would have me carted off to a psychiatric unit. If Lane slowed up a bit, a call of "Go Lane!" would spur her on to run more and faster. It would also prolong the pleasure of our having a dog that was glad to see us every time we returned.

Lane wanted to go with me everywhere I went, but sometimes even I wasn't expecting her to follow. On Whitney's wedding day, I was still working until the very last minute on some basic projects to finish the chapel I had built for her wedding. When time ran out, Lane and I went to the house to get dressed for the ceremony. Whitney's Aunt Kathy had made a big white bow and tied it around Lane's neck. The bow dressed Lane up for the occasion, but the white of the bow brought out the white of Lane's graying muzzle. Sometimes I teased Lane and said, "Come on old woman," when she was moving slowly. She lay on her bed with that sad look that said, "You're putting on a tuxedo. I know I'm going to be left at home today." So, she was surprised when the mule drawn wedding carriage arrived and I called her to go with me. I helped her into the carriage and we rode side-by-side down to the inn to pick up Whitney, where she was dressing in Magnolia's cottage. I helped the beautiful bride aboard the carriage and the three of us rode to the chapel.

The photographer was with us every step of the way to record our ride and arrival. We greeted the rest of the wedding party and Lane got compliments on how pretty she looked with her bow. While everyone waited for the processional to begin, Lane patrolled the chapel grounds to make sure no uninvited birds, rabbits, or squirrels had invaded the premises. I lost track of Lane as one by one the bridesmaids made her way into the chapel.

Finally, everyone was inside except Whitney and me. As the bridal march began playing, the bride and father of the bride stepped through the open door. I felt something press against my right knee and glanced down to see Lane coming in with us. With Whitney on the left and Lane on the right, the three of us walked down the aisle together. It looked so orderly that the guests thought it was all planned. No one was more surprised than Whitney and I. The groom must have been thinking the father of the bride might give away the dog. Not a chance!

To Lane Two Too

When I first saw her, she gave me a look that I should have recognized, but I didn't. She was two months old and the last in the litter to be claimed. Everything about her was puppy, except the look she gave me. Her soft bushy puppy fuzz made her appear to be a ball of fur with legs. She was more body than legs, but the legs kept her in constant motion, as she scampered around the living room floor. She was learning to retrieve a ball. I guess it was a good thing for a Golden Retriever to be learning. When the ball rolled in my direction, I retrieved it before she could get to it and tossed it across the room. She bunny-hopped after it like she continued to do when she got excited. She reached out with a paw like it was a hand and stopped the ball's roll, then grabbed it with her mouth like it was a squirrel about to get away. She pivoted and came back in my direction. I thought to myself, "this is a smart dog who already knows a lot." I reached out to accept the ball, but when she was about four feet away, she stopped, froze, and gave me the look. I didn't know it at the moment, but I was going to get that look many times in the future. One of her first puppy pictures with her new bright red collar captured the look. It was the look that said, "I know what you mean and I know what you want, but I don't want to do it. Not only that, you can't change my mind or make me do it."

It had been nine months since Lane, our first Golden Retriever, had died, and we were looking for another one. Well, Claudia, was looking for one. I was along for the ride. I didn't think it was the right time to buy another dog. It wasn't that I felt more time should pass in order to get over the loss of the other dog. It was winter and I didn't think that was a good time of year to get a pup. How do you convince a pup that it is better to go outside into the cold and learn to pee on the grass instead of staying warm inside and peeing on the plush carpet? They just stand at the open door and give you the look.

But here we were an hour from home in Culleoka, looking for another red Golden Retriever. We had been happy with the rich copper color of our first one, even though guests at our inn would ask, "Is she an Irish Setter?" I've responded a hundred times, "No, she's a red Golden Retriever." "I thought Golden Retrievers were blond." "Well they can be blond, gold, red, and even white."

A deal was quickly struck. As we headed home, we tried to think of names. I suggested we name her Lane, the same name as our last dog. She had become the spokesdog for our inn and was famous for her newsletter column and website features. Giving this dog the same name would allow that role to continue seamlessly. After all, the next bear doesn't get a new name when Smokey dies. The University of Georgia doesn't try to think of a new name when their mascot, Uga, is replaced. Claudia countered that this dog would be different from our last one and should have a different name. She didn't realize how prophetic she was in saying this dog would be different. Maybe she had seen the look.

As the fur ball snoozed in my lap, we drove along exploring possible names. Since we had been brilliantly creative in naming the last dog after the Lane family who raised her, we tried on Smith for this pup's family. "No way! How about Culleoka? We just bought her in Culleoka." "Are you serious? That's no name for a dog unless it's an Indian." "Well, how about Culley? They were the first settlers on our farm." "That would make a good name, but it sounds like a boy's name. We've got a girl." The trip and the name game continued as the pup slept through it all. By the time we pulled into our driveway, it was settled. Claudia reluctantly gave in and we carried Lane into our warm living room for her first pee on the plush carpet.

We called her Lane, but sometimes the conversation included, "Which Lane?" So, we came up with Lane I, Lane II, and even Aunt Lane. Later, when we started giving her a daily pill for her allergies, we called her Zyrtec at pill time. There was no problem with her taking the pill. I would just hold it up in the air and she would sit down and wait for it to be put down her throat. I was glad I didn't get the look. The pill had to be put all the way to the back of her mouth. Even without the look she would walk away and spit out the pill. I don't think she liked being called Zyrtec any more than swallowing the pill.

As Lane grew, I waited for her puppy fuzz to disappear, but didn't notice much difference. Instead of developing long, coarse hair like Lane I, this dog's hair was still like baby hair – short, fine, and soft. On a visit with the vet, I asked if Lane still had her puppy hair and the vet said, "This is the kind of hair she will always have. What you have here is a red, short-haired Golden Retriever." The red, golden, and retriever descriptions weren't a surprise to me, but short-haired was. I had never even heard of a short-haired retriever, but that's what we had. It's a trait we grew to like as much as her rich red color. The inn guests were fooled by the short hair too and would comment to Lane, "Oh look, you've had a haircut." Lane would just give them the look.

We admired the long wispy plumes of feather-like hair on the traditional Golden Retriever's tail and hindquarters. They looked like golden thread when the sun caught them waving in the wind. However, we didn't like finding the clumps of bushy hair that fall out when traditional Golden Retrievers are shedding their winter coat.

We weren't surprised at Lane's intelligence like we were with Lane I. With no Golden Retriever experience, we were way behind the curve on training Lane I. She was showing things to us that we hadn't taught her, but she knew. I don't think we every caught up. With this new dog, we were determined to get an early start on training. Lane II was already retrieving the ball, but unfortunately, she had developed the look.

When she was a few months old, I was sitting on the couch, and she came over and sat in front of me. That was no great feat, but Lane didn't just sit there with the tip of her tongue showing through a slightly open mouth giving me that Luck's Bean dog smile. She reared back to sit straight up onto her rear end and held her front paws out to balance herself. I first thought she was about to jump up onto me, which would have been a bad habit to start. But she just sat there as if to say, "Let's talk." It was always her favorite way to greet guests. They aren't sure what Lane is trying to do and while they might expect a short-bodied dog to balance itself like this, they are amazed at this long lanky-bodied dog doing it. Like when I first saw her rear back, sit down on her rear, and reach up with her paws, they expect her to jump up onto them next. "Just hold my paws a minute and let's get acquainted," Lane seems to say. "Everyone else is grabbing hands, why not me too?" It's no wonder that dogs look a bit embarrassed when someone asks them to shake hands. Anyone would be embarrassed if someone asked them to shake hands and they didn't have a hand, but not Lane II!

When Lane was six months old, we left her for a few days with our daughter Whitney in the mountains of North Carolina, while we went on to the Outer Banks for a vacation. Whitney introduced Lane to the mountain streams and couldn't wait until we returned to show us how Lane would run down the stream dipping her head into the water every few feet. The more the water splashed, the better she liked it. After several minutes of this vigorous exercise, she would need to rest and would lie down in the water to cool off. We don't have any mountain streams at home, but when we have a big rain, she pretended the ditches were mountain streams and romp through them.

Whitney took Lane to the annual Montreat Fourth of July Parade, on a leash of course. There would be a thousand people, lots of kids, lots of dogs, and picnic on the grounds. Lane did fine with all the people, kids, and even had her own picnic. But she was a parade stopper, - literally. Lane decided to investigate something interesting and was stopped abruptly at the end of the leash. She dug in her heels when Whitney pulled hard on the leash to get her to come back. Lane didn't want to come back, so Whitney pulled harder on the leash and when she did, Lane's collar slipped over her head and off she galloped. There were no commands to get her to stop. By the time Whitney caught up with her, Lane had joined the parade. With no collar to grab onto, Whitney just tackled Lane and pinned her to the ground. Not wanting to go back, Lane just rolled over onto her back, laid there, and gave Whitney the look. The parade waited until the leash was retrieved and Lane was coaxed away. Whitney was embarrassed as several people gave them looks of their own. Lane watched the remainder of the festivities from the confines of the car.

When we started teaching Lane verbal commands, I remembered one of our guests had a dog that responded to backwards commands. They didn't sound backwards to the dog, just to people. If the dog was standing nearby and heard the command "Come", it would lie down. I know, that sounds like what all of our dogs did when we were trying to teach them to come. And when we tried to teach them to stay, they would run away. But this dog was doing exactly what the owner wanted it to do. On the command "Down", it would sit-up. On the word "Up", it would come. The dog didn't know the true meaning of the words. It had just been trained to hear the words with a different meaning. It made me think of how dogs around the world respond to words. A Frenchman wouldn't praise his dog by saying, "Good dog!" He would say, "Bon Chien!" The poodle would just give me the look if I told it "Good dog!" and Lane wouldn't understand "Bon Chien", that is unless I taught her. So that's what I did. She learned to recognize words in six languages. She knew French, German, Italian, English, and Spanish, and we worked on the Chinese command "Fu Cha" for body down.

Like a child, the first command I needed for Lane was "No." I decided on the German "Nein." As a pup, Lane heard Nein a lot. Nein! Don't chew on the furniture. Nein! Don't jump on me. Nein! Don't bite my hand. I soon found this different language idea had a downside. Guests at the inn would say "No" to her and she would just give them the look. This time the blank look meant "I don't have a clue what you just babbled." I also discovered I wasn't the only one using different commands to train a dog. Police dogs and other working dogs aren't trained to the traditional sit, stay, come commands. Imagine a German Shepherd charging a burglar only to have the burglar call off the attack with a simple shout of "Sit!" They probably don't even recognize Nein and they should know German, shouldn't they? On the other hand, I would instinctively say "Nein" to someone else's dog and it would give me the look. Actually, I guess this wasn't officially the look, since it meant, "I don't know what you mean, but I'll be happy to do what you want, if you will just speak in plain English."

Lane's next word came quickly on the heels of "Nein." Lane and I were in the breakfast room. I was getting a fresh dose of puppy breath, while she washed my face with her slick little tongue. (I've since developed new thoughts about this face licking. It was after watching a dog lick its master's face, when the moment before it had been giving itself a butt licking!) Claudia was getting ice from the in-door dispenser to make a gin and tonic. Lane heard the strange new sound of the rumbling refrigerator and went to investigate. "What's that sound about?" Lane asked. It wasn't the moment to teach Gin Time, that would come later. But it was time to teach Ice. While Lane chased the elusive slippery ice cube around the kitchen floor, Claudia and I tried to think of a language to use for ice. Unfortunately, the word ice in most languages is pronounced just like the English pronunciation. We finally cheated and settled on the Italian "glace" for ice cream.

I got a grin from people when I wanted Lane to sit and I said "Park It!" Spanish speaking people whip their heads around to see what's going on when I call out "Andale" to signal to Lane it's time to "Hurry up, let's get going!"

My favorite time to use foreign words with Lane was at the farmer's co-op. Lane was always ready for a ride in the truck to any destination. Even if I'm only moving it from one spot in the driveway to another, she had her feelings hurt if she's not invited to go on the twenty-foot ride. I stuck with the standard English for having her go to the truck. I yelled "Truck!" and she would go to the cab and wait for me to open the door. But, if I wanted her to jump into the truck bed, I yelled "Guterwagen!" in German.

I was standing on the co-op loading dock and talking with some local farmers while the workers loaded my feed. Lane was roaming around the warehouse checking out the smorgasbord of smells. There were only two familiar smells for Lane. One was the bags of dog food and the other was the warehouse cats. Lane didn't spot the cats on her first trip, but she did find where their food bowls were kept and cleaned them out like eating a bowl of candy. Nothing tastes quite as good as someone else's food, even if it's a dog eating cat food.

I was anticipating and relishing the moment when the loading was complete and it was time to call Lane, because I knew I would get a reaction from the farmers. I shouted "Guterwagen!" and Lane came running out of the warehouse like she was about to be left behind to live with cats forever. She jumped onto the truck and stood on top of the feed sacks. The farmers listened and watched with astonished looks on their faces. Finally, one of them asked, "What did you say to that dog?" I only replied by saying, "Truck" and Lane, hearing the cue, jumped down off the feed to go stand by the cab door. "That's some dog you have there," one farmer replied. I only opened the truck door and said "Arriba!" Lane hopped onto the truck seat and we drove away leaving the farmers standing on the loading dock. I can only imagine the conversation they had after we left.

Adults have to be careful around children, because the children will start using words, that they overhear the adults using. It may be cute to hear a three-year-old say, "I want a cookie, damn it!," but they will also say it when the minister is visiting. The same care needs to be taken with pups. They are constantly listening and observing. Unfortunately, like children, they learn words and attach them to the wrong things. We never have figured out how this happened with Lane and the words "Mashed Potatoes." We only eat mashed potatoes about twice a year, but one evening Claudia called from the kitchen to ask, "Would you like mashed potatoes for dinner?" Lane had been relaxing on the rug in front of the couch, but hearing the question, she jumped up like a fire alarm had sounded. At first, she gave a low growl and went to the glass door for an anxious look outside. Then with a bark, she went from window to window and back to the door, looking for an invader. We had no idea what was going on. There surely wasn't a giant potato about to attack us. Claudia said, "Mashed potato?" again and Lane's agitation increased. The only time we had seen this behavior before was when Lane would look outside and see our neighbor's dog, Rose, in our yard or on the deck. Rose lives a mile away, but would occasionally stop by on her neighborhood rounds and visit with Lane They were good buddies, but Lane always greeted Rose with a growl to remind her whose territory this was. In ten years, I had never heard Rose bark at anything, even once. Lane and Rose gave each other the ritual butt sniff to make sure no one is an imposter. (If dogs have such a great sense of smell, why do they have to put their nose an inch away from another dog's butt to smell it?)

Had Lane come to associate the words Mashed Potatoes with Rose? If so, how? They don't sound anything alike. Mashed Rose? Rose Potatoes? We couldn't figure it out, but the next time we saw Rose outside, we called out "Mashed Potatoes!" and Lane sprang into action. If there hadn't been a connection before, there was now. We even started calling Rose "Mashed Potatoes" from time to time and told our neighbor about Rose's new nickname. When they used the nickname around Lane, they thought Lane's growl meant she didn't like Rose. I could be far from home in downtown Chicago and say "Mashed Potatoes!" to Lane and she would start looking for Rose. People don't believe it, but they change their minds when I reluctantly demonstrate it. It's not really fair to stir up Lane needlessly. Around the house, we just say "M P" when we're planning the menu. We don't think Lane can spell – yet!

Most of Lane's adventures were in her first year. She passed over the traditional eating my shoes and went straight for the arm of the love seat in the breakfast room. She got to the heart of things in a matter of minutes and was pulling the stuffing out. By the time we discovered the situation, she was down to the wood frame. No strategically placed quilt would cover this damage. We were destined for a new love seat. Claudia gave Lane the "Nein! Bad dog!" rebuke. Lane gave Claudia the look.

On a long weekend trip out of town, we left Lane at our inn for the staff to watch her. We put her food and water bowls in my office along with the familiar blanket for her bed at night. During the day she was able to run freely inside and outside the inn. When she turned up missing one afternoon, it was all hands-on-deck to look for her. The innkeeper called some of the neighbors with no luck. The housekeeper looked all over the property and checked with guests for clues. The maintenance man drove up and down the highway looking, but there were no signs of Lane. The next morning calls were made to area vets to put out the word on the missing pup. When the innkeeper gave Lane's description to a local animal shelter, they said they may have our dog and we should come to town and claim her. They advised the staff to bring $49.50 to pay for the overnight stay. (At that price, she could have stayed at Day's Inn with room service!) The staff took money from the petty cash box and bailed Lane out of jail. The officer told them there was no clue about Lane's owner, since there was no ID on Lane's collar. He also told them Lane had come to one of our neighbor's house and, since Lane was new to the neighborhood, they didn't know where she had come from. When we returned from our trip, the staff was relieved to report a story with a happy ending.

Before we could get an ID for Lane's collar, she repeated her disappearing act. This time we were at home. I figured I could go around the neighborhood calling for her and she would come to me. I walked about a mile in every direction calling for her. I called "Lane!" and "Andale!" and even her favorite word "Truck!" As the sun set, I gave up the search and walked back home. The next morning, I went door to door to inquire with the neighbors. By this time the neighbors were probably wondering what was going on with the Hazelwoods and their new dog. One of the neighbors said they had locked a dog up in their horse trailer to keep it away from their horse colts. They said it was the same dog that animal control had picked up a few weeks ago and they were going to call for another pickup that morning. Lane was happy to see me and happy to be released from the trailer. I was happy to not get the look and happy to have avoided another $49.50 in bail money. Lane had heard me calling, but she couldn't escape the trailer to come to me. I was too far away to hear her, if she had barked.

If Lane could talk, she would have a lisp. When she was about a year old, she got excited over catching a ball and bit her tongue. When I first saw the situation, I had no idea what was happening. She walked up and gave me that half open mouth smile, but her nose, mouth, and face were covered with blood. Blood was running freely from her mouth. At that moment all I could think of was she had an internal hemorrhage and was spitting up blood. A quick examination revealed she had torn a notch the size of my thumbnail out of her tongue. The lost notch was nowhere in sight. It was just as well; there would be no repair of this wound and the healing would be slow. Somehow the bleeding stopped, and she seemed unconcerned about her wound. In time the wound healed with help from Mother Nature. The only consequence was I noticed a little water flying off to one side of her mouth when she drank from her bowl. Sometimes when she came up to me after a run, she had her long tongue hanging out about six inches for all to see the old war wound. I told people she got the injury in a fight with a wild animal and left it at that. When I called her by her Indian name, Tongue Notch, she gave me the look. We gave Lane I an Indian name, Tongue Spot, to go with the ink-like spot that appeared on her tongue, but she never gave us the look.

The notched tongue didn't slow down or impair any of Lane's lickings. She maintained her coat like a cat, by constantly grooming herself. She may even have given someone's hand a lick as they held her paws when she reared back to sit on her rear. It would be a small gentle lick that meant she liked them or at least liked what they ate with their fingers last.

We were glad to see Lane give Addison, our new granddaughter, an "I like you" lick on the head after sniffing her bald head. I think her mom, Whitney, was happy about the friendly introduction, but could have done without the lick Lane gave her infant daughter. Whitney would have really objected if I had brought up the butt licking subject.

Lane and Addison grew to be fast friends, especially since Lane let Addison do almost anything to her. A hard pat that felt like a slap, a tug on the ears, or a finger poke in the eye were each received cordially. If Lane couldn't take it anymore, she just got up and left Addison sitting in the floor.

Plans for raising a litter of pups with Lane were still on Claudia's mind. Unfortunately, Lane came into heat for the first time, just before we left for twelve days in Spain. The staff was alerted to the situation and advised to keep a close eye on Lane. We suggested it would be best to not let her run loose. When she was outside, it would be best for her to be on a leash. When we returned, they assured us they had watched Lane closely, but a few weeks revealed they hadn't been one hundred percent effective. Lane was pregnant. The sire was unknown.

When the pups were born, it was a mixed litter of half black and half red. The black color pointed toward a neighbor two miles away who had a Rottweiler. At weaning time Claudia posted a video on the Internet in which she described each pup. It was during basketball season and she had given each pup the name of a Vanderbilt basketball player. With a few emails to spread the word, people flocked to the puppy box in the corner of her Vanderbilt office. They were there to celebrate Puppy Day and maybe even claim a player-pup, three-point shooter or rebounder for themselves.

With the responsibilities of motherhood behind her for now, Lane resumed her responsibilities at the inn. She picked up where Aunt Lane left off as the Director of Public Relations. Many guests weren't aware there had been a change of dogs as they called Lane by name when they arrived at the inn. "Hi Sweetie," they might say as she greeted them in the foyer. I would be out of sight around the corner at my library desk and could hear all of this. I would respond with "Hi, how are you?" They would reply with "Oh, we weren't talking to you," as they rubbed Lane's upturned belly. Lane was always the first to be spoken to. She continued to handle all of the communication with the advisory board to assure it would get read. She also continued her monthly column in the newsletter and on the website.

While Lane I spent the day in her club chair in my office, Lane II camped out at my feet under the desk. If someone came for an appointment and I was ready for them to give their attention to me instead of Lane, I said to Lane, "Go to your office." Office to her means it is time to get under my desk and she made her way around and under the desk. Sometimes her presence continued to be revealed by a tail sticking out from under the front of the desk. A mention of her name created a thumping of her tail on the floor to amuse the guests further.

Lane enjoyed having the run of three hundred acres around the farm, but she wouldn't be considered a farm dog. She was a business executive. She couldn't be counted on to be helpful with the cows. She liked to chase the cows, but there was a better than fifty-fifty chance she would chase them at the wrong time or in the wrong direction. She liked to get nose to nose with them and see who flinches first. She always won because she knew how to give them the look and they turned and ran away in defeat. She was counted to keep the cows away from the fence. If they wandered too close to her territory, she sent them away with a charge and a bark.

I could usually count on Lane to be well-behaved. Her manners were so good that when we were at home and sat down at the table for a meal, within one minute she would go stand by the door, asking to go outside, so we could eat in privacy. She expected the same privacy when she ate.

She was a good traveler when we took her with us on a long car trip. Lane I wanted to watch the highway every minute in case she needed to find her own way back home. Lane II may have watched for a minute, but then she would lie down in the back seat and snoozed the whole trip.

When she rode shotgun in the passenger seat, she took her position with her right elbow leaning on the arm rest. If we were driving slowly through town with the windows down, she may have hung her head out for fresh air and catch the attention of the people in the car beside us. "Who's your girlfriend," they might yell. "I like red heads!" was my reply.

Living on a farm gave Lane an advantage over city dogs. A city dog has to depend upon its owner to bring a chew bone or save one from a pot roast at supper. When Lane wanted a new chew bone, all she had to do was walk up the hill to the animal bone yard and take her pick. The bone yard was a dormant sink hole about six feet deep where dead cows could be left to decay without smelling up the place. After a while, nothing was left but a pile of bones. Sometimes Lane brought back a cow's femur that can handle months of gnawing with little effect. Lane usually got bored with it and put it under a bush for a while. The next trip she may have come back with a jaw bone with all the teeth still intact. The bones that got the biggest reaction were the entire skulls fit for a Georgia O'Keefe desert painting. The kids thought the bones were from dinosaurs, and much to the chagrin of their parents, I encouraged them to take one home. After all, Lane knew where to find plenty more.

Lane's cow bones don't get to come inside the house. She has a wicker basket of city dog chew toys and stuffed animals that are kept in the kitchen under the butcher block. I like to watch her go to the basket and search for the one that suits her mood at the moment. We haven't taught her to return them to the basket when she's finished. I've noticed it takes a while to train kids to do the same thing.

Another training that was always in process was her approved locations for sleeping. During the day, when we were at home, she never attempted to get onto our bed or the sofas. But, if Claudia or I was having trouble sleeping in the middle of the night and moved to a sofa, we had to be careful not to lie down on top of a dog. A nudge was not enough to get her up. There was never a reaction. Either she was dead to the world or being stubborn about wanting to move. She kept her eyes closed, so we didn't get the look. It took a drag by the legs to pull her off the sofa and onto the floor with a thud. Propping the cushions on their edge for a few nights broke the routine and she usually stayed in her bed all night.

She showed no interest in our bed, except when we left her alone at home. When we got back, she was not on the bed, but we could see the bedspread had been wrinkled and knew she had been on it. Lane would be standing nearby with a sheepish look of "I've not done anything." An inquiring hand to the wrinkled spread revealed a warm spot that told the true story. To keep the bedspread clean, we had to resort to putting her blanket on our bed, knowing she would be up there before our taillights were out of sight. It might not have even take that long. One day I needed to go outside to the shop for a tool and, since it was raining, I didn't want Lane to go out and get wet. I told her to "Stay" and she gave me the "I'm sad I can't go" look. I was gone about two minutes, and when I returned, she was on the bed. Maybe she thinks "Stay" means get onto the bed.

When our inn guests complimented Lane's good looks, friendly disposition, and well-mannered behavior, I usually bragged on her some more. Then I told them, she was all I ever dreamed for in a girlfriend. She wanted to be with me all the time and go wherever I went. She always wanted to be close and liked to be touched. Once, after telling this, a man said, "Yes, I understand she has all you want in a girlfriend – bad breath and yellow teeth!" I countered with, "well, she doesn't have yellow teeth, but I admit the bone yard does give her bad breath from time to time." Lane was insulted by the comment and gave him the look.

Upside Down Horse

Livestock are a curious bunch. When something new appears on their horizon, they may stand still and stare at it for several minutes without moving to see if it is a predator. They will also stare at an inanimate object, like a different piece of equipment parked in the field. The expression, "like a calf looking at a new gate", is used to describe someone who sees something new, but won't move even if it is a way out. But after a time of watching with no movement, livestock will gradually move closer to investigate. They will watch, listen, smell, and even stretch their neck out long enough to touch it with their nose. If still nothing happens, they may give it a lick to examine its flavor.

Sometimes their curiosity gets them into a difficult situation. How they react to being trapped depends upon the species. Cows will first struggle to escape, but if not successful in a few minutes, they become resigned to being held and just stand there waiting to be released. Horses, on the other hand go berserk, when trapped and will struggle until they injure themselves or even die.

This was the dilemma facing Charlie, our Tennessee Walking Horse. I say he was ours, but he was really ours in a partnership. My wife and our friend, Rich, struck a deal to get another horse. Rich didn't have a horse or a place to keep one, but enjoyed coming to our farm and riding our horse. We had plenty of pasture on our farm and my wife thought it would be nice to have a second horse, so we could go for rides together. Rich offered to buy the horse, if we would provide pasture in the summer and hay in the winter. Here's a piece of advice. If someone makes you a similar offer, you should agree to cough up the big bucks on the front end to buy the animal and let them feed it. You will come out ahead that way.

Charlie was a mature, well trained gelding that never gave us a problem. That is until one weekend, when I was out of town and Rich was left in charge of the farm. Now, Rich grew up in the city and will tell you he learned all he knows about the farm from being at our place. Other than a cow getting out through the fence, according to Rich, there's seldom anything to go wrong. (That's a lie!)

Charlie's curiosity got him into a situation, that created a problem for Charlie and Rich. Knowing he was going to ride during the weekend, Rich put Charlie into the barn lot Friday morning. There was water available, a big tree for shade, and ample grass for the day. When Rich came home from work that afternoon, he noticed Charlie didn't come to the fence to greet him, expecting the usual treat. A quick scan of the barn lot didn't reveal a horse so Rich went to the barn to see if Charlie was loafing in the shed. He wasn't there either. Where could he be? It wasn't likely that he jumped the fence. Had he been stolen? Maybe he was lying down on the other side of the hay stack and couldn't be seen. Horses don't lie down very often. They even sleep standing up. So, when I see a horse all stretched out in the pasture, I am afraid it is dead.

Our hay was baled into large round rolls that weighed over twelve hundred pounds each. I would stack them end to end in a couple of rows until winter, when they would be picked up with the tractor and taken out into the pasture one at a time for feeding the cattle and horses. Rich looked on the other side of the rows of hay, but Charlie wasn't there either. He was headed back to the house to get out of his suit and tie when he heard a muffled grunt from between the two rows of hay. A couple of feet were left between the rows, so the rolls wouldn't be touching and air flow would keep the hay from spoiling. When Rich looked down the seventy-five-foot row, he saw Charlie about half way down the row lying on his back with all four legs sticking straight up. Charlie couldn't roll left or right because of the rolls of hay on each side. There wasn't even an inch between the hay and Charlie's heaving sides. The weather was cool, but in his panic and struggle, Charlie had worked up a sweat. He tried to raise his head, but it could only come up a foot or two. The flailing of his legs in the air had even less effect. Charlie's eyes were as wide with fear as they would have been if a mountain lion was attacking him. Rich's eyes weren't as big as Charlie's, but he was sympathetic because he was afraid that he couldn't get Charlie out of this predicament before he died.

How did he get into such a fix? How long had he been like this? From the looks of the ruffled hay on the sides of the rolls, he had been kicking at them for some time. He probably went between the rows of hay out of curiosity to see what was there or mindlessly grazed a step at a time down the row. He could have just walked out either end but probably suddenly realized he was trapped on two sides and panicked. He was afraid to move forward deeper into the row, couldn't turn his long body around, and tried to back out. Walking backwards is a very unnatural movement for a horse, so he probably tried to turn around and fell over onto his back. Regardless of how he got this way, he was like a turtle laying on his back and waiting for someone to turn him over onto his feet again.

Seeing Rich there to rescue him wasn't any relief. He still struggled in terror like his life was at stake. Maybe it was. Rich wasn't feeling relieved at finally finding him either, because he had no idea how he was going to get Charlie unstuck. Even if he could have picked up Charlie's head and pivoted it toward his hind legs, Charlie wasn't about to let him do it. The rolls were too heavy to push out of the way and the tractor was away being repaired. That left Rich with only one solution. He cut the strings on a hay roll and used a pitch fork to pull hay away from the roll a little bit at a time. The baler had wrapped the hay tightly into the roll so it was slow going. With Charlie still struggling and kicking constantly, Rich had to be careful not to jab the horse with the pitch fork. There was a lot of hay to fork out. Would Rich get Charlie free before he had a heart attack or died from exhaustion? Rich could sympathize, because his heart was pounding with terror and he was exhausted too.

Rich finally got enough hay removed on one side that he thought he could free Charlie, if he could get him to roll in that direction. Unfortunately, Charlie wasn't trained like a dog to roll over on command. He pulled on Charlie's halter to get his head moving in the direction of the hole he had cleared. Charlie heaved himself toward the hole and twisted the front of his body, but fell back onto his back because he couldn't get his hind feet turned and under him to get up. Horses are just the opposite of cows, when they get up from lying down. A horse gets up front feet first, and cows get up hind feet first. Charlie still needed to get his hind feet under him to get up and out.

Rich started digging at the hay in the next roll that was beside Charlie's hind feet. It was slower work, because he had to dodge the flailing hooves from Charlie's hind legs. That's all Rich needed was for his wife to come home and find a trapped horse and a knocked-out husband. After getting about as much hay from the second roll as he had the first, Rich wiped the sweat out of his eyes and grabbed one of Charlie's hind hooves as it came by. Before he could pull on the leg to roll Charlie to one side, Charlie kicked the leg out of Rich's hand and sent Rich sprawling. Charlie may have been near exhaustion, but he still had more power in his hind quarters than Rich could hold onto. That wasn't saying a lot because Rich's arms were as limp as a dish rag by this time.

Maybe if Rich got the horse's head turned again, there would be enough room for Charlie to turn the back of his body on his own. Rich crawled over the nest of hay he had created and up to Charlie's head. If this didn't work, it was back to clearing away hay a handful at a time. Rich pulled on the halter; and as Charlie's head came forward, the horse's terror-filled eyes were just inches away from Rich's. Hot breath was blowing from the horse's flared nostrils. With a clamber, Charlie turned his front body enough to get his front feet under him and, with a strain like an old man getting up out of a chair, Charlie slowly stood up.

Rich and Charlie just stood there looking at the situation. Both were weak in their knees and content to just stand still for moment. Charlie shook himself and Rich took him by the halter to lead him down the narrow alley between the rows of hay. Rich contemplated how he was going to explain this experience. He didn't know who was the most relieved, him or Charlie.

Sunshine on a Wedding Carriage

It was an unusually warm September Saturday morning and the bright sunshine made it feel more like summer than fall. Claudia and I decided it would be a perfect day for an outing to an auction in the community. We weren't going to buy anything. We were just looking. Now everyone knows how that turned out.

This was not an ordinary estate auction. A man in our community, who made horse-drawn wagons, buggies, and carriages, was having an inventory reduction auction. I always thought inventory reduction was the goal of every sale, but it seems to get used frequently as a come-on in advertising.

Claudia and I were making plans for Whitney's June wedding and had thought having a horse and carriage would add grandeur to the occasion. Most families rent a carriage, but since we were also building a wedding chapel for her, we thought we may as well buy the carriage too. This sounds a bit over the top, but we were already hosting weddings at our inn. The chapel and carriage would make useful additions. We had never shopped for a horse and carriage, so we had no idea what we were looking for or what they would cost. Besides this was a day for shopping, not buying. Right!

After a five-mile drive down one country road and then another, we came to a surprisingly large crowd of people milling around about fifteen buggies. We didn't know the interest in buggies was so big and wide. Every style of buggy one can imagine was on display. There was a black one-seat doctor's buggy, a farm wagon painted John Deere green and yellow, a six-seat surrey with top (no fringe), even a miniature farm wagon with two ponies hitched to it. Kids huddled around the ponies and adults crawled in and out of buggies. If a horse was hitched to a buggy, some took them for a test drive down the county road.

After we looked at the surrey with an old gray mare attached as a candidate for our wedding carriage, we saw a gleaming white vis-à-vis carriage pulled by a trim young sorrel mare coming back down the road and into the auction lot. It had two maroon velvet seats facing each other. (I later learned that's what vis-à-vis meant.) The top which covered the passenger seats could be folded back on a pretty day like today, but it was up to give the passengers a shady ride. There was a single seat up front for the driver with a holder for a buggy whip at his side. (Why are they called buggy whips? Who whips a buggy?) Thoughts of the surrey evaporated. Now this was a real wedding carriage.

It was time for a test drive. The owner introduced us to the mare and said her name was Sunshine and she was a registered Tennessee Walking Horse. His wife had raised her as a colt. She had never been ridden under saddle but had a month of training to pull a carriage. He assured me that she had no bad habits. I climbed into the driver's seat and with a shake of the lines eased her out of the lot and down the road. After about two hundred yards of slowly clip-clopping along, we turned around and headed back to the auction lot. I heard a car coming from behind to pass us. This would be a test for Sunshine. How would she react to car traffic? She had been calm with all of the people milling around her at the auction lot, but a car coming from behind would be different. With the blinders on her bridle, she wouldn't be able to see the car until it was almost in front of her. The car passed us and Sunshine never flinched. That was a good test and Sunshine passed it.

With a hundred yards to go, I shook the lines and she moved into a smooth-running walk with no hint of wanting to go even faster, breaking into a canter or galloping away. I pulled her up as we neared the lot and walked her over to where the mother of the bride stood. "How did she do?" Claudia asked. Sunshine had passed all of the tests I knew to give her.

After a $3,500 bid we were the owners of a carriage. And what good is a carriage without a horse? Our choice came down to the two-year-old, inexperienced, sorrel mare and a ten-year-old gray mare once owned by an Amish family. Since the Amish mare may not have had many years left in her, we went for youth over experience. For another $1,500 we had a complete outfit of horse, harness, and carriage. It wasn't a bad morning for a couple who were just looking.

Now how would we get our purchases home? I could go home and get our horse trailer to haul Sunshine, but I didn't have a trailer big enough to haul the carriage. Since we now owned a carriage horse, we decided we may as well use her for the purpose we bought her. It was only five miles home and at a walk that would only take an hour. At a running walk it would be less than that.

The only part of the plan that made me nervous was the one mile that was along U.S. Highway 41-A. It had a lot of traffic moving at high speeds, but it had a very wide, paved right-of-way and I could drive the carriage on it without actually being on the highway.

With Claudia driving our truck behind us with its emergency flashers on, we headed toward home. I shook the lines and Sunshine eased into a running walk with her head straight ahead. It didn't take very long for my first surprise. There was a white feed sack in the ditch up ahead and I noticed Sunshine keeping her curious eyes on it as we approached. Gradually her head kept turning to keep the sack under surveillance. The sack was a new sight for her and she didn't know what it might do. As we got almost even with the sack, Sunshine jumped to the left to give the sack more distance in case it attacked her. Her veering to the left surprised me and I quickly pulled on the right line to get her back on track. I was glad there wasn't a car coming when she jumped into that side of the road. Almost as quickly as she had broken stride, she settled back into an easy running walk with her head straight ahead.

No sooner had I gotten relaxed from the sack episode when I heard a loud engine ahead. I knew from the sound it was a big truck, even before I saw it pop over the hill ahead. It was not only big; it was a semi-trailer truck loaded with logs. It wasn't its size that worried me though. It was its noise. Sunshine had earlier passed the test of a car coming from behind, but it just purred as it passed. This truck was roaring as it came down the hill toward us. After the sudden swerve Sunshine had made passing the sack, I was worried what she might do at the sight and sound of this big log truck. The truck driver didn't help me any. As he came down the hill, he pushed on the gas and made the noise even louder. I gripped the lines even tighter, sensing the unexpected and feeling out of control and out of options. This was all up to young, inexperienced Sunshine. At that moment I wished I was behind the old gray mare, but Sunshine never flinched or broke stride as the truck roared past. As I relaxed my grip on the lines, I realized

I had stopped breathing and was tense all over. Listening to the clip clop and watching Sunshine pull us effortlessly straight down the road helped me relax.

After a mile down the county road, we came to the intersection of the highway that worried me the most. I turned Sunshine onto the right-of-way of highway 41-A and she ignored the traffic coming both ways just as she had ignored the log truck. I later learned none of this was new to her, since part of her training had been alongside this highway. I would have been more relaxed, if Sunshine had been Mr. Ed and told me she had been on this highway before.

To keep this part of the trip as short as possible, there would be no slow walking until we went a mile to the next county road. Even though the first half mile was up hill. I kept urging Sunshine to stay in a running walk. I began to see some drops of sweat break through her shiny coat as the heat of the day and the hard work took effect.

When we approached the top of the hill, Sunshine's head eased to the right as her gaze spotted something on the roadside ahead. I didn't see anything moving like a dog or wild animal, nor did I see another contrasting color like the feed sack. Whatever it was, Sunshine was focused on it just like she had been on the feed sack. This was no place to be swerving suddenly into the highway. The traffic was fast and constant. I eased Sunshine as far to the right of the right-of-way as I could without getting off into the grass. I still couldn't see anything remarkable on the roadside ahead. Suddenly Sunshine came alongside her potential predator. With her head turned to one side to keep an eye on it, she swerved to the left again. The extra space I had created was enough to keep us from swerving into the lane of traffic. And what was the predator? A five-foot gray boulder was protruding from the green weeds of the roadside hill. The contrast of the green and gray was enough to seem ominous to Sunshine. With another predator averted, Sunshine resumed her sweaty running walk to the top of the hill.

Reaching the top, Sunshine and I both needed some relaxation so I pulled her into a slow walk. She would also get to relax as the carriage pushed her down the hill. I just hoped the steel shoes on her feet wouldn't slip on the asphalt like skates on ice as we went down the hill. The weight of the carriage quickened her pace back into a running walk. With a left turn across the highway approaching at the bottom of the hill, it was time to try the carriage brake for the first time. Sunshine and I were both relieved as the brake did its job of easing the load and slowing our pace.

I began watching traffic from both directions. Would we be lucky enough to have a break in the traffic, so we could make our left turn without stopping and before another speeding car popped over the hill? I didn't want to have to sit on the side of the road and wait our turn to cross. On a glance behind me I noticed Claudia had the truck's left turn signal flashing. This would at least give notice to any traffic. Luck was with us when we got to the intersection. There was no traffic in sight and we made our left turn without stopping.

I knew the next mile of this tree-lined country road offered just what Sunshine and I needed. It was made for a nostalgic carriage ride. The trees came together over the road to make a shady tunnel. Through the clip-clop of Sunshine's hooves I could hear the water running in the roadside creek. Its water added to the coolness of the shade. As refreshing as the water would be, Sunshine would have to wait for the end of our trip to get a drink.

The last three miles were uneventful, as we went from one country road to another. We were even spared a train whistle as we crossed the railroad bridge next to our farm. The carriage brakes lightened the load going down the hills, but there was no way for Sunshine to escape the hard pulls up the long hills. Her coat was wet all over with sweat when we turned into our driveway. It may have been a strange new place to Sunshine, but I felt our adventure had ended.

In front of our house I pulled Sunshine to a stop, which she was quite willing to do. As Claudia and I talked about the trip home, Sunshine stood motionless with her head down as far as the harness would let it drop.

Claudia brought a bucket of water and I removed Sunshine's bridle, so she could lower her head and drink from the bucket on the ground. That was a big mistake – removing the bridle, not giving her a drink. With the bridle blinders removed and without ever taking a drink, Sunshine turned her head and saw that big white thing that had been following her. She bolted forward to escape the carriage before I could grab onto her head. Off she ran at a gallop into the yard with the driverless carriage chasing her from behind. As she went under the hackberry tree at the side of the house, the carriage top caught on the lower limbs and was pulled back, but Sunshine kept galloping. As she made the turn to disappear behind the house, the carriage rocked over onto two wheels and back down again. As I ran around to the other side of the house to head her off, I had no idea what might emerge from behind the house. I didn't know how I would stop her or catch her, but I ran anyway. I got to the other side of the house just in time to see her and the carriage still in one piece. She was still in a gallop and headed toward a cherry tree and maple tree at the corner of the yard. It would be a miracle if she maneuvered between the trees. There was no miracle. The hub of a front wheel gashed into the trunk of the cherry tree. That's where the carriage stopped abruptly, but Sunshine kept galloping. Her harness had broken free from the carriage and she was now free to escape. She hardly broke stride, left our yard and crossed the driveway into an unfenced forty-acre hay field beside our house. Parts of her broken harness flopped and fell off as she left a trail of leather in her wake. How far would she run? I hoped she would stop before she got to the road. I watched her gallop at full speed with an occasional look over her shoulder to see if that big white thing was still chasing her. After about two hundred yards, she came to a stop about thirty yards from the road. She turned and looked back my way with her head high, ears perked forward, eyes wide open. With a snort she shook her head, as if to celebrate her escape.

Claudia inspected the carriage and tree damage, while I got a lead rope from the stable and went to retrieve the horse. Sunshine didn't move as I made my way toward her, but I knew she could bolt at any moment. I was a stranger to her with nothing appealing like a bucket of feed, so I wasn't sure she would let me approach her without running to escape. She might blame me for this whole episode, which would be appropriate.

She watched me cautiously as I approached her slowly. I never hesitated as I moved forward and eased right up to her. Some horses let people walk up to them but run away when they reach for its head, but Sunshine didn't move when I reached for her. With her bridle laying in the driveway at the house, there was nothing for me to grab hold of but her mane, which I did. After a rub of her neck, I eased the lead rope around her neck as I spoke to her softly. Sunshine was probably the only word she understood, so I was talking to calm myself as much as her. With a close hold on the lead rope, we walked the harness-strewn path back to the house with me glad it wasn't wedding day.

Fortunately, I had nine months to do harness and carriage repairs and additional training for Sunshine. I wasn't going to be surprised, if Sunshine and the cherry tree would be scarred for life. Only the metal frame of the carriage top was broken and a few hundred dollars took care of it. The local harness maker made the new, but broken harness look new again. After a month with the trainer, which included some driving time with me, Sunshine showed no ill effects from our mishap.

The trainer brought Sunshine home, confident she was ready to pull the carriage again. She seemed a little nervous when we pulled the carriage shafts onto each side of her, but she relaxed while the trainer and I attached the repaired harness. I suggested that he be the driver and Claudia and I would enjoy the ride. My confidence hadn't fully returned.

He drove us the half mile from our house down the hill to the chapel site construction. The only hesitation was when Sunshine crossed the wooden bridge. The change in surfaces from asphalt to wood made a different sound with her hooves. When she looked to the right and saw the running water down below in the creek, I thought she might swerve, but she looked to the left and saw the same sight and decided straight-ahead-with-a-quickened-pace was the best route.

We crossed the bridge on the return trip and the trainer stopped the carriage and offered to let me drive the rest of the way back up the hill to the house. Thinking of few obstacles ahead, I moved into the driver's seat; and the trainer sat in the back with Claudia. We got about half way up the hill and I noticed it was hard for Sunshine to get good traction on the driveway, so I pulled her to one side into the grass. By this time, she had lost her momentum and came to a stop. The trainer suggested I give Sunshine a touch of the buggy whip. I gave her a light flick and she leaned into the harness, but getting the carriage going again from a dead stop on this hill was going to take a harder pull. "Give her some more whip," the trainer called. When I did, Sunshine reacted immediately, but not with a harder pull. She had already decided she didn't want to pull the carriage up this hill. She reared up onto her hind legs, lost her balance, and fell to the ground onto her side, breaking a carriage shaft in the process.

She didn't thrash about trying to get up or get away as I expected. When she didn't move at all, I was afraid she was injured and jumped from the driver's seat to attend to her. Her legs were tangled in the broken harness, so if she was able to get up, she wasn't going to gallop away again. When I got to her head, she had a terrified look in her eyes. I didn't know if it was from fear or pain. Among the broken shaft and harness, she rolled over onto her stomach and tried to stand, front feet first and then her hind feet. The fall must have just stunned her or knocked the wind out of her because she was apparently uninjured.

Once again, I had a broken carriage, broken harness, traumatized horse, and uncertain owners who were glad it wasn't wedding day. It was embarrassing to return to the carriage and harness makers once again needing repairs. We waited until spring to return Sunshine to the trainer for repair to her confidence. Claudia and I spent the winter working on our confidence too.

I didn't think it would be wise to take this young, inexperienced horse into the crowd of wedding guests. All kinds of images popped into my mind. None of them were pleasant. I hadn't even thought far enough ahead to know who would drive the carriage. I guess I had always assumed it would be the father of the bride. I didn't think I wanted to be responsible for a real-life runaway bride.

Sunshine did just fine with her training, but the trainer said he thought at two years old she was too small to pull the carriage, especially up our hills. I had evidence to think he was right. He thought pairing her with an older, more experienced horse would keep her calm and lighten the load.

Borrowing the old gray mare came to mind, but her larger size would make them a mismatched pair. The trainer was a mule man and didn't have a horse to use, but offered a pair of his mules. I would have to talk with Claudia and Whitney about that idea. I'm sure it wasn't what we pictured in our dreams of a horse-drawn wedding carriage. I knew the trainer could be our driver, but his John Deere cap and overalls weren't in our picture either.

On wedding day, the carriage looked like new. It had been repaired and showed no signs of its mishaps. A neighbor had made a cloth banner with the words "Just Hitched" which we tied to the back of the carriage. Sunshine was unfazed by all of the excitement. She watched every move of the crowd from her paddock atop the hill at our house. The two mules looked as grand as mules can look, but not nearly as grand as our trainer in his black suit and the top hat we rented for him.

As attached as we were to Sunshine, her favorite pal was our Golden Retriever, Lane. I would catch them nose to nose, exchanging sniffs. I even saw Lane stand for Sunshine to lick her back. Apparently, Sunshine liked the taste and Lane liked the massage and grooming. These two red friends were a sight to watch as they raced around the paddock. Lane was sometimes chasing Sunshine from behind and trying to grab her tail. Then they exchanged places and Sunshine chased Lane ending with much rearing and pawing that could have led to injury, but they both knew it was all for fun.

As the years passed, Sunshine was never hitched to a carriage, buggy, or wagon again. She did gallop free through our yard and hay field many more times. First, I let her loose in the yard to nibble grass. She would eat grass that came up in Claudia's flower garden, but, unfortunately, she liked to pull the hostas out of the ground too. Rather than go back to the stable for water, she refreshed herself with a long drink from the bird bath.

In the winter she made her way to the garden for some sweet clover that was planted for a cover crop. By spring she was going to the hay field where she could eat her fill. It was a pretty sight seeing the afternoon sun shine on Sunshine's red coat as she galloped free in the green hay field. At sunset I could whistle and she would gallop across the hay field back into the stable.

Dead Skunk in the Middle of the Road

Which smells worse, a skunk because it's dead on the road or a skunk because it smells like a skunk? Regardless, it's a horrible smell. Our family has a song we sing when we are driving down the road and come upon this familiar odor. "Dead skunk in the middle of the road. Dead skunk in the middle of the road. Stinkinnnnnnnnnnn to high heaven!" (Sung to the tune of "Dead Skunk in the Middle of the Road.") By the time the song is over, the smell has usually dissipated unless we were unfortunate and hit the skunk with our tires. In that case the smell could follow us for days. Valet parking really doesn't like us showing up like this. The drive-through banker doesn't welcome it either.

My grandfather, Ollie, was throwing hay from the loft to the cows and saw what first looked like a black and white barn cat. It was in the loft looking for a nest the chickens used to hide out and lay their eggs. It only took a split second for Ollie to realize it wasn't a cat, but a skunk. That split second wasn't long enough because the skunk sprayed Ollie before he could move away even one step. This wasn't just a little passing spray; it was the full load, direct hit, scent gland emptying, clothes saturating kind of spray. It blinded Ollie and choked him so much that he could hardly find his way out of the loft. There was no need to get away from the skunk. It had already done all the damage it could do.

Ollie made his way to the house, wiping his face with a handkerchief. When he stepped from back porch into the utility room my grandmother, Irene, chased him back outside. "I got sprayed by a skunk," he said. "You don't have to tell me. I can smell it!" she shouted.

The old remedy for getting rid of a skunk spray was to wash oneself, the clothes, the dog, and whatever else had been sprayed with tomato juice. It would take all of the tomato juice Irene had canned from the garden that year to wash Ollie and his clothes. She decided Ollie was worth the washing, but the clothes weren't.

"You take all your clothes off, put them in the trash barrel, and set a match to them," she said with no hesitations. Ollie went out behind the smokehouse, stood beside the trash barrel, and stripped off every stitch of his clothes and threw them into the barrel. As he ran naked back to the house, he had two wishes. He hoped no one pulled into the driveway at that moment. And since he could still hardly stand to smell himself, he hoped Irene would let him back into the house. The skunk spray wasn't in his hair, where it would have lingered a week or two. Ollie was as bald as a cue ball.

I don't know how many dogs I have had in my life, but that's how many have been sprayed by a skunk. I would look out into the yard and see one of my dogs rolling in the grass and rubbing its head with its paws. I would go out to look at the dog and to see what was going on. I didn't have to get all the way to the dog to get my answer. The answer came floating my way. It was another skunk-spraying victim.

I've never been sprayed by a skunk and don't want to be, but there were a couple of times I thought I would get sprayed. The first was on the third tee at a golf course. After hitting my ball, I turned to go back to my cart and saw a momma skunk with six babies come walking around the cart in my direction. I decided she could have the cart, if she wanted it, but she was headed to the woods where she probably lived. I gave her a wide berth, knowing it would be easy to upset a momma skunk protecting six little ones. She waddled off into the woods and out of sight. Even if I hit the ball into those woods on future rounds, I would just let it be a lost ball rather than taking a chance on stumbling onto a skunk. Now that would be a golf hazard!

The second time was at a friend's house and I was walking around the house from the front yard to the back. Without paying much attention to where I was going, I walked right upon a skunk. I hightailed it back to the front of the house before the skunk could do his own hightailing and spray me. When I told my friend about the skunk, he just laughed. It turns out it was his pet skunk and had been de-scented. I've never understood why someone would want a pet skunk. Wouldn't a black and white cat do just as well? To me, a skunk just looks like it smells bad. I've also wondered how the vet is able to do the surgery without getting sprayed.

The last time my dog was sprayed, I decided to call the vet and ask for a recommendation on what to use to get rid of this smell.

The vet said they had a product called Skunk Away, but it wasn't much more effective than the old tomato juice remedy. The last time we used tomato juice on our dog, it jumped out of the bath tub, went into the bedroom, jumped onto the bed, and rolled. Not only did our bedspread smell like a skunk, it had a pink stain from the tomato juice. I bought a wash tub for the next washing outside on the lawn. The vet's recommendation was to use vinegar. The acetic acid in the vinegar acted on the skunk spray in the same way as the tomato juice. Plus, it was cheaper and a lot easier to rinse away. The vet said, "To tell the truth, nothing works very well."

We keep a gallon jug of vinegar in the garage just waiting for the next inevitable skunk spraying. I think I'll go now and label the jug "Ollie's Skunk Antidote." The dog that smells like a pickle is probably mine, but it's a lot better than the alternative of stinkinnnnnnnnnn to high heaven!

The Author

David was reared in Kentucky, but his writing may take you down many of the roads he has traveled in his roles as consultant, farmer, minister, restaurateur, innkeeper, and pencil seller. His list, "Jobs I've Had," tallies thirty-four, and he's unemployed (most call it retired) and looking for work now! He confesses that at the core of all these endeavors is his desire to help things grow—first himself, then plants, animals, people, and organizations.

Beginning in the ninth grade, he told everyone he was going to be a veterinarian. That response continued until he was introduced to organic chemistry in college. After his prayers for help went unanswered, he attended seminary and became a campus minister.

His degrees in animal science, theology, and business administration all came together to help things grow on a three-hundred-acre farm as innkeeper at Parish Patch Farm & Inn, Cortner Mill Restaurant, and Whitney Wedding Chapel in Normandy, Tennessee.

Animals, church life, and a priority for family have been constants in his life. They continue to nourish his life and writing as he and his wife, Claudia, live with their Golden Retriever, Lane III, thirty cows, three hens, and a goose under a pear tree.

Let's talk…
david @wufu3.com
www.wufu3.com

More from David B. Hazelwood

Cortner Mill: A Community Treasure

Our restaurant didn't have secret recipes. If you liked what we served and wanted the recipe, it was yours for the asking. But, if you want all 240 of the recipes in our cookbook, you will have to pay for them! We will throw in some history of our 1825 grist mill and stories about our recipes for free.

Miss Lizzie's Heirlooms

In the early 1900s my grandmother started writing down her recipes with a pencil in a blank page memorandum book. Her forty-nine recipes are the core of this recipe book along with stories about her cooking for a family of ten during the Great depression. Aunts, uncles, and cousins have added their recipes and memories of eating from Miss Lizzie's kitchen.

Lists for Ainsley

Inspired by a list of favorite things by Nashville sportswriter, Fred Russell, I started making lists of my favorites: things I don't do anymore, sweet things to do for your spouse, pet peeves, and scores of others. They range from sublime to ridiculous and will provoke you to think of what's on your own lists. The book is not finished yet because I keep thinking of more lists to include!

Monday Morning Rose

I sent a rose to Claudia after our first date and continued for more than twenty-five years. Each was delivered to her with a personal note of love from me. Sixty-four of them were selected for this book from more than a thousand she saved. We didn't include the X-rated ones.

More to come…

My Short Life as a Step Dad

Whitney was seven when I married her mother. We know I became her step dad on our wedding day, but somewhere through a lot of meaningful experiences I became just dad. Like me, you will laugh and cry your way through the many episodes I've recalled and maybe you will discover how to become just a dad too.

Cooking Southern: Recipes and Their History

This is much more than a collection recipes of the historic South. It explores why Southerners eat what they eat and why they use their methods of cooking. Southern food came to us from enslaved Africans, European immigrants, native Americans, and hard-pressed Southerners coping with civil war. As you explore this book's more than 1800 historic recipes, you can create your own version of Cooking Southern.

Made in the USA
Columbia, SC
08 February 2022